PENNY PORTER

ADOBE SECRETS

ADOBE SECRETS

by Penny Porter

Wyatt-MacKenzie Publishing, Inc.
DEADWOOD, OREGON

Adobe Secrets by Penny Porter

Illustrations by Marilu Savage.

Previously publshed in 2006 by Singing Valley Press
Original ISBN - 0-9656923-3-7

S E C O N D E D I T I O N

ISBN: 978-0-9743832-3-1
Library of Congress Control Number: 2008942317

W

Wyatt-MacKenzie Publishing, Inc.
D E A D W O O D , O R E G O N

Wyatt-MacKenzie Publishing, Inc., Deadwood, OR
www.WyMacPublishing.com (541) 964-3314

Requests for permission or further information should be addressed to:
Wyatt-MacKenzie Publishing, 15115 Highway 36, Deadwood, Oregon 97430

Printed in the United States of America.

Contents

Dedication

I dedicate this little book to my children, God's most precious gift. Emmy, Bud, Jennifer, Scott, Becky and Jaymee, with hope that when the going gets rough, they, too, can escape into the joyous world of memories like their mom.

THIS OLD HOUSE

An adobe hut, a century old,
Has a history   needing to be told.
Instinct for shelter, mankind's third,
Likewise   for animal or bird,

Enchants the   mud brick walls of Earth,
Where drifters find an inner worth,
Plus self respect in mein and mood
And confidence in attitude.

Where little children redefine
What Nature, man and myth combine.
So love and simple virtues win
In ways that make you cry and grin.

The stories Penny tells are true,
And bless whatever shelters you.

Cassius Sargent: Poet

Introduction

The short story is great bedtime reading, and my purpose in putting this little dream catcher together was to include just enough stories in a book that the reader, instead of being overwhelmed, would want to read slowly, all the stories, one at a time—perhaps one each night—depending on how weary he or she might be. Also, I wanted to keep the size of this collection small enough to fit into a bedside table drawer instead of the table-top from where it might slip, only to land under the bed and hide until discovered by the vacuum sweeper that slides beneath low places.

It is said real people are more fascinating than fiction. The characters and the animals in this book lived. They include a drifter, a Hell's Angel, a soul with a dark past, an illegal alien, a "witch," a bird, a bull — each one dwelled at one time in the crumbling adobe shack on our ranch. From where they came is not important.

But what we learned is. We watched. We listened. They each had a story to tell. And in the telling there was a re-living cloaked in secrecy that brought life to a time that spanned one hundred years or more — and each left a message of love.

Penny Porter

THE LADYBUG MAN

On the northwest corner of our desert ranch where dust devils dance and yucca bells chime, a crumbling adobe bakes like a muffin in the searing Arizona sun. Its mud-brick walls are split and scarred by time. Chunks of mortar and shards of purple glass skirt the old foundation, and the weather-beaten door sags like a broken jaw until the wind coaxes its rusted hinges to creak and groan in dismal harmony with tattered wires and a dented stovepipe that clack like castanets across the corrugated metal roof.

And to think that someone once lived here?

I muse over nearby Tombstone's legendary gunfights, deaths and disease in the adobe's early days. The mining town was home to early gold and silver seekers who lived in clustered, tin-roofed huts much like this one. Some came in wagons bringing women and children, pots and pans. Some brought Bibles and dogs. Sometimes one brought a bird in a cage. Friendships were born. Nights kindled dreams. Days nurtured hope.

But why did one homesteader choose to build so far from the rest? Nobody wants to live alone on the desert, away from the comfort of voices. There are times when the knot of loneliness tightens the heartstrings of even a twentieth-century rancher's wife, and I find myself wishing the adobe held a neighbor with whom to enjoy a memory, tell a story, or maybe share a dream.

Who *did* live here nearly a century ago?

I shade my eyes and peek through finger slivers of light at the withered desert. That's when I see him – a puff of dust, a phantom creature shrouded in the shimmering mirage hugging the edge of the long dirt road; a man so thin, he's almost not there at all. Yet something deep inside me whispers...

"It's him! It's him! He's coming home."

• • •

So it happened the following afternoon that our two little girls burst into the kitchen like Irish leprechauns, ponytails green with alfalfa. "Mama! Mama!" Becky squealed. "The ladybug man fixed the door on the tumbled-down house!"

"The ladybug man?"

"Our new neighbor," piped four-year-old Jaymee, "and he's going to paint his mailbox so it will be ready when his letters come."

Indeed, two mailboxes shared a common T-post at the crossroads: "Ours," the box with RR 14-A painted on the front, and "theirs," the box that belongs to the old adobe, the box rusted shut—the ugly box. Yet, on early summer mornings when I stroll up the coyote-worn trail to put letters in RR 14-A, I see beauty there, dewdrops dripping from the spider web spun the day before that joins *our* box—and *theirs*.

After lunch, I filled a plastic milk bottle with cold water and packed a turkey sandwich, a few chocolate chip cookies and a shiny red apple in a brown paper bag. Feeling pied piper–like, I walked the powdery path with two children singing "Ladybug, ladybug, fly away home..." plus three dogs, four barn cats, and assorted chickens intent on the contents of the bag strung out behind me.

"Why do you call him the Ladybug Man?" I asked.

"Because he's got one," said Becky.

"Where?"

"Under his hat."

There was no time to pursue the conversation. We

approached a stick figure, easily ninety years old, wearing a New York Giants baseball cap and spraying a hinge with a mushroom cloud of WD-40, undoubtedly retrieved from my husband's workshop. He smiled a golden tooth, his only tooth, and I knew he couldn't bite into the apple. He tipped his hat too fast. I missed the ladybug.

"You must be our new neighbor. Welcome... Mr...?"

"Name's JJ, ma'am." His hand felt like a turkey claw in mine. "I lived in this house... a long time ago. Me and my mother. And Canario."

I gave him the bag, wishing I'd put more in it. JJ waved at the interior of the house.

"Yep. I'm gonna fix 'er all up... already chopped a window in the wall to let the sunshine in."

I glanced at a hole near the roof where four blocks had been hacked away. "Why so high?" I asked. "You can't see out."

A faraway look clouded his eyes. "Ah," he murmured, "but I can watch the birds fly by."

My gaze shifted to two orange crates wedged in a corner. One served as a shelf for his few possessions, including a wallet, a tin coffee cup and a harmonica. The second bowed under the weight of stacked drawing paper, a wide-mouthed mason jar stuffed with red, green, and blue pencils and crayons, and a sharpening knife. Colorful drawings of birds papered one wall from ceiling to floor. All were signed "JJ."

A closer look at the sketches revealed a sameness; against a blue background, every bird, wings catching the sun, seemed to dash rays of golden light down, down onto the solitary figure huddled in the lower right-hand corner.

"You're an artist, JJ?"

"Yes, ma'am." He flashed his 14-karat grin. "I used to bring in a bundle of money with my oils and watercolors—when I was young."

I wanted to question him, but for now, there were more important matters. "There's a shower and toilet attached to the horse barn, an old army cot and blankets in the storage room, and a chair, and a little potbellied stove we can haul up here in my pickup."

"That'd be real nice, ma'am."

From that day on, JJ spent early morning hours sweeping my chicken coop, gathering eggs and raking the barnyard. Come noon, I often spotted the baseball cap disappearing among the manzanita bushes and clumps of bear grass, and I couldn't help wondering where he was off to. But it was the hot afternoons that touched my heart, when I frequently caught this elf-like old man and my two little girls enjoying lemonade at the picnic table in the shade of our ancient cottonwood. There he taught them how to draw birds. And they taught him other A, B, C's besides J and J.

When evenings came, sweet harmonica notes wafted from the old adobe, filling the prairie nights with birdsong. He wasn't what I'd had in mind, but I had a new neighbor.

• • •

"The Ladybug Man still never gets any mail, Mama," Becky said. "His box is always empty."

"Does he send any?" I asked.

"We don't think he can write." She frowned. "But I'm teaching him."

Jaymee squashed an ant with her tiny sneaker. "He can't read either."

"Well, why don't you girls surprise him? Send him a letter!"

JJ's happiness over that first letter in his mailbox and others that followed unlocked his extraordinary gift of storytelling, chapters from a very long life. Tears pooled in his rheumy old eyes when he folded the letters and tucked them back in the envelopes. Magically, they triggered memories of a different world where he was once a rich and famous artist and an art collector.

His stories were outrageous, unbelievable, but as the days and weeks passed we all fell under JJ's spell.

"Got a letter today from my old friend, Rube," he began, caressing the envelope swelling his shirt pocket. The children exchanged silent giggles. "Got my first Porsche when he bought four of my paintings. That was forty thousand smackeroos!" On another day, "Got a great letter from Mario." He smiled. "We owned the biggest casino in Atlantic City—my paintings all over the walls. He bought them all." And yet another, "Can't recall which movie I starred in but they filmed it at the Metropolitan Art Museum." The details that followed spun around the wealth he once enjoyed from the sales of his "masterpieces" that hung in galleries all over the country. He had a new story every time the girls managed to sneak another letter into his mailbox. And we loved them all.

We paid JJ for chores so he had money to spend with Dan, a mysterious friend who arrived in his '57 Chevy on the second Tuesday of every month to take him "shopping for supplies in Tucson." We could hear the old wreck coming, its sagging muffler grinding into every pothole on the 20-mile stretch of dirt road from Tombstone. When it echoed its return in the wee morning hours, I knew JJ was safely home—the kind of home that belonged to the desert, to time, and to him.

• • •

Not a week went by that JJ didn't find quiet hours to add a new drawing to his already cluttered walls. The composition never changed. Birds flew by a high open window with background in blue, and sunbeams poured onto a small thin figure seated on the floor in the lower right-hand corner.

• • •

One day, the children discovered a new litter of calico kittens among towering bales of alfalfa in the hay barn. A letter for JJ's mailbox was forgotten. Since their dad was working on his tractor nearby, I decided to search for arrowheads and pottery chippings frequently exposed after a heavy rain in the deep arroyo that sliced through our ranch.

Shadows were lengthening when I heard a child's voice above the eroded, root-tangled slope. I scaled the bank for a better look and peered over the edge. There I saw JJ, on his knees, rocking, praying, beside what appeared to be a grave, his baseball cap pressed to his heart.

"JJ! What are you doing way out here?"

"Talking to Mama." He didn't seem to recognize me. I scrambled over the top to kneel and put my arm around his shoulders.

"It's okay, JJ. It's only me."

"I have to tell her about Canario—tell her I'm sorry."

"Canario? Who's Canario?"

"Mama's canary." He paused for a minute. "Mama loved him so. He sang so pretty in his little wooden cage. Then she got the fever and they wouldn't let us in town. Papa built the house and she got worse every day. Finally he left and said he'd be back with medicine, but he never came. He took my shoes so I would stay put. But we needed water."

I was afraid to speak. Tears pooled in his eyes and coursed through the deep wrinkles mapping his face. "I was only five years old, but Mama needed that water bad—and I knew there was plenty in the White River Draw."

"That's a four-mile walk!" I gasped, envisioning the trek through miles of dense cholla, prickly pear and mesquite to get there. "How could a small child carry water that far without shoes?"

"I did it, but only once. I got that water for Mama and me. I poured it on her lips, in her mouth. We needed more, but my feet were so swollen I couldn't walk any more. Then, early one morning, Mama opened her eyes and whispered. 'Canario? Johnny. I can't hear him singing. Did you give him water?'

"'No, Mama. I saved it for you—and me.' I remember climbing onto the table and peering down at the tiny, yellow-feathered body on the bottom of the cage. Canario was dead. It was my fault. And Mama stopped breathing—before I could say, 'I'm sorry.'

"Men came, but they were afraid of the smallpox. They dragged her off with horses and lassos and I followed them to this place. Every day after that I put on Mama's shoes and carried rocks to cover her grave and keep her safe from the wolves and coyotes."

"But who took care of you? What did you eat?"

"Once in a while someone came and left food inside our old wagon. I peeked out the door and they yelled at me, 'Stay in there till we go!' Then, one day a Mexican family came. They took me out of my house, away from my mama, but they were good to me. I had brothers and sisters until I was twelve, but I ate too much. It was time to run."

I glanced at the sky. "We need to head home, JJ," I said. "It's getting dark." He followed me back to his house, took off his cap again and thanked me. That's when I saw the little red birthmark with tiny black dots on top of his head. *The Ladybug man.*

"I'll be leaving for the East with Dan in the morning, ma'am. Will you look after my mother?"

"Of course I will."

"I need to go check in with family," he added.

Family? Not once had JJ ever mentioned family.

• • •

Two months later, a letter came from Miami Beach, Florida. *Dear folks,* it began. *Thank you for taking care*

19

of my grandfather. I'm sure he told you he spent 50 years in prison. I have three small children, so please don't think bad of me that I couldn't take him in when he got out. He arrived at our local hospital a week ago, sat down in the waiting room and died. He was sort of crazy and fancied he was a great artist. I enclose a drawing that was found in his backpack. He was 96 years old. The letter was unsigned. With trembling hands I unfolded the drawing.

Perched in a high window, background cobalt blue, a tiny yellow canary raised its head in song. "Canario," of course. Canario! I fought the ache in my throat as my eyes tracked the golden rays of sunshine spilling onto the too-thin man seated on the floor of what I now knew to be a prison cell. The initials, JJ, were where they always were, in the lower right-hand corner. But this time other letters had been added. A-U-D-U-B-O-N.

J. J. Audubon. Of course! John James Audubon, the idol of so many young artists in the nineteenth century. One above all. My neighbor — *The Ladybug Man.*

Carefully, I put the picture back into the envelope, and instead of taking the coyote trail home, I walked past the old adobe and headed west up the arroyo, west to a lonely grave in the desert where a wounded little boy carried rocks day after day to cover his mother and keep her safe. I will never go to this place again, but on that day I had reason; a need to bury a masterpiece of a small yellow bird under one of those rocks, and hope that somehow, somewhere, it would bring song once again to a woman long forgotten in Arizona history, and a little boy who never had the chance to say, "I'm sorry."

AN ANGEL IN CHAINS

Cheeks flushed with excitement, our four-year-old daughter climbed over the fence of the corral where I was bottle-feeding an orphan calf, and squealed, "Mama! Mama! You have to come see! Angels wear chains!"

I was about to ask, *Becky, what on earth do you mean?* when my heart plunged to the pit of my stomach.

Outside the wooden gate amid the cactus and mesquite towered a stranger, his skin gleaming like oiled mahogany in the blistering Arizona sun. Nearly eight feet tall in giant-sized motorcycle boots, he wore a red sweatband that failed to control the black braids leaping wildly from his head. In the early morning breeze he seemed to sway like a genie uncorked from desert sand.

Heavily muscled arms stained with purple tattoos burst from a leather vest. A deep scar crimped his left cheek and a small silver dagger swung from one ear. But it was the chains on his boots, chains on his belt and chains cascading down his massive chest that made me wonder,

Why does Becky think he's an angel?

Then I spied the Harley Davidson parked in the shade of the old adobe shack at the end of our long dirt road. On the far side of the barbed-wire fence a gang of wind-whipped, grease-streaked, smoke-shrouded motorcyclists milled around, "HELL'S ANGELS" glinting across the shoulders of more than one black jacket.

This was 1974. I'd read about the Hell's Angels—terror-riddled tales of large groups of men who adhered to no boundaries of human decency, infamous for murder, rape, theft, guns and drugs. My husband was at a bull sale in Casa Grande. The children and I were alone. Why was this group hanging around here on our ranch a hundred miles from Tucson?

"His name's Rip because his muscles ripple," Becky piped, as she ducked beneath the fence this time and took the stranger by the hand. "His motorcycle broke." She tugged him toward me, and although he seemed bigger with each step, I noticed he looked down at my little girl as though seeking reassurance.

Finally, he bowed his head and his uneasy, hooded eyes met mine. "Rip Balou, missus. I know it's gettin' late, but two of my buddies took off for the city to get me a new clutch. They won't get back till morning and I wondered if we could camp near the shack and water-pump for the night? We won't bother you none... and... all we need is water."

Dared I say no? It was a chance I had to take, yet something beyond Rip Balou's frightening appearance—and the Hell's Angels' reputation—made it seem safe to say, "Sure." I glanced at the group by the gate. "But please," I said, "don't smoke. Fire danger is at a peak right now."

"Don't you worry yourself none about no fire," Rip said. The thought seemed to have humbled him. "Those warning signs are posted all the way from New Mexico." He thanked me before walking back toward his friends.

"But Mama, what about supper?" Becky asked. "They don't have any food."

"How many are there?"

"Eleven... no, nine. Zack and Ty went to Tucson... and Rip makes ten." She answered so quickly. *Zack?...Ty?...Rip?* I wondered how long she'd been down by the gate. Long enough to count—to know their names—and to make a friend! Such a natural thing for a child to do, especially one without playmates. I vowed to keep a closer eye on her, but at that moment my thoughts were on food. I fed everyone else who stopped by—ranchers, cowboys and Mexican *mehados* hoping for work. What harm could possibly come from feeding a band of ... Angels?

Later, back at the house, ten men sat at the picnic table under the cottonwood tree drinking iced tea from Styrofoam cups while Becky held them spellbound with a Barbie-doll fashion show. As they wolfed down tacos and beans, I asked questions. "Where are you going?" Los Angeles, and they had been on the road for two years, ever since they had met at the Harley Davidson Rally in Sturgis, South Dakota. Before that time, some had come from major cities across the country. Chicago. New Orleans. Boston. New York. "And home and family?" I asked. Few responded, but Rip, the obvious leader, muttered "Baltimore, 2,647 miles away."

The following morning, Rip's huge frame darkened the kitchen doorway. He didn't look happy. "Zack's back. They had to order the clutch from Phoenix," he said. "It'll take a coupla days. Could we stay? We could rake... clean stalls... do somethin' to help out?"

"Okay. I guess you can't get very far without a clutch." I thought I was being funny. He didn't.

"And missus. There's 12 of us now." I knew he was referring to meals.

Soon, more hands than I would ever need, or find again, unloaded a double semitrailer-load of hay, repaired fence, and rode back and forth to Tucson to buy food that I hadn't even asked for. I noticed they laughed and talked a lot among themselves. Why not? I thought. No responsibilities. No family ties.

Strangely, it was big Rip Balou who not only worked the hardest but continued to be drawn to Becky—and she to him. She let him help bottle-feed the orphan calf and collect eggs in a basket from the chicken coop where the ceiling was so low he couldn't stand up straight. Then, when she placed a day-old chick in his enormous hands, his mouth opened like a child who had just touched Santa.

Three meals a day at the picnic table left time to share more than her Barbie doll. Although not yet in school, Becky could read, and I watched a remarkable friendship tighten between her and the giant man as they hovered over a book together. Was it possible that a little girl could make a difference in his life?

Rip's tough, big-shot countenance seemed to soften, and the face of a once small boy emerged. In his eyes I saw a whole life flash by as Becky ran a tiny finger beneath magic words that introduced an Angel to... *Beauty and the Beast*. Rip watched. He listened. I wondered... could he read?

What did it matter? It was Becky's crayons and coloring books that caused those haunted eyes to brighten. "Red and blue are my favorite colors," she told him, "but we can share. Can't we?"

It wasn't long before the crayons in his pie-sized hands created magic of their own. Rip banished another Angel to Tucson to "buy more." During the two days that followed, he taught Becky how to coax pastels from primary colors and fill empty skies with sunrises, sunsets and rainbows. Gradually, every page in the coloring books became a Rip Balou masterpiece.

"I don't like ugly, dark colors," Rip told Becky. "Anyone can color like that." Then over and over again, he covered her small white hand with his huge dark one, and said, "Honey, the most important thing to remember is—you gotta to stay *inside* the lines."

It was after supper on day three that Becky popped the question. "Do you have a mommy and a daddy?" Rip didn't answer. Instead, he flexed his muscles so the ship on one arm seemed to roll in a storm and the dragon on the other coiled to strike. But he'd shown her those wonders before. Now there was something else on her mind. She asked him again.

Reluctantly, Rip unhooked a leather pouch from the chain around his waist and pulled out a photograph of a gray-haired woman with glasses. Her hand rested gently on the shoulder of a little girl. "That's my mama," he said, "... and that's Jasmine... my baby. She'd be just about your age now."

"I wish she could come play with me," Becky said.

Rip stared at the picture for a long time. "Mama's raisin' her," he said, "but she's got the glaucoma. She can't see so good no more."

Becky fixed her eyes on Rip then, and in the infinite wisdom of a child she asked, "If your mama can't see so good, who's going to teach Jasmine to stay inside the lines?"

Rip shook his head. "I... don't... know." He answered softly, but I could hear the pain in his voice, his heart and his soul.

Late Thursday evening, the gang members who had gone in quest of the new clutch finally returned. Rip must have worked throughout the night to get his bike running, for they were all ready to leave at daybreak. Although the barnyard and corrals had been raked and the men were filling their canteens with cool water, I sensed unrest among them. "No breakfast, ma'am," one Angel

26

said. "We gotta hit the road." He glared at Rip. Had there been an argument? *A disagreement?*

Careful not to scare the horses and chickens, their bike motors purring softly, the Hell's Angels one at a time cruised over to say "good-bye" and "thank you." Rip Balou was last.

"Thank you, missus," he murmured, "... for a lot of things."

"And thank you for being such a wonderful play-mate—and teacher," I answered. I wanted to say more, ask him why he had chosen such a life, but suddenly his eyes were brimming with tears over which he had little control.

"I don't see Becky," he said, glancing over at the pic-nic table. "I... I need to tell her somethin'... remind her of somethin' real important..."

"She's down by the gate." I hugged him quickly, straightened the chains around his neck and found my-self wishing I could do the same to the chains that had stolen his life. I pointed to a very little girl sitting alone on top of the fence. By now Becky was waving and shout-ing good-bye above the roar of impatient engines as one by one the Angels turned *west* on Frontier Road—*west* to Los Angeles—leaving the peaceful desert of moments ago drowned in swirling dust. "You better hurry, Rip," I urged. "They're going to leave you behind."

He smiled at me then—for the first time—a big smile before coasting down to Becky on the high ground be-tween ruts worn by tires. I watched him set the kick-stand before he dismounted and walked over to the little girl he'd grown to love. He lifted her off the fence and set

27

her down on the leather seat. Then, crouching beside her so they could speak face to face, Beauty and the Beast talked... about sunrises?... sunsets?... rainbows?... Who knows?

What I do know is that the good-bye hug he gave Becky brought tears to my eyes. Then the last Angel swung a long leg over his gleaming Harley, revved up the engine and turned *east* on Frontier Road—*east* to Baltimore – "2,647 miles away," where a little girl waited for a lesson on the importance of staying inside the lines, and a daddy to show her how.

GOLIATH'S SECRET LIFE

Cold March winds stampeded across the flatlands of our Arizona Hereford ranch. Every time the barometer fell, cows calved, and so far we'd had thirteen heifers. "Where are the bull calves?" my husband, Bill, kept asking. It was 2:00 a.m., and my job was to check the herd just in case a pregnant cow was in trouble.

On this particular night, as I headed toward the calving pasture on my all-terrain motorcycle, the wild bobbing of the headlight seemed to propel me into an eerie world. Ghostly feline shadows popped in and out between fence posts. Wild shapes slithered along barbed-wire fences, and silver-trunked mesquites and cactus, familiar in daylight, became pulsating, alien creatures until, at last, the herd of white-faced cattle welcomed me, their eyes aglow like kitchen lights of far off farms and ranches.

They watched me, unafraid as I drove among them. Some were lying down with babies huddled close to their sides. Others stood bunched for warmth, rumps to the wind. But my grip tightened on the handlebars when I spied one cow standing alone in a distant corner of the field, her nose sweeping the ground, nostrils distended, back humped, her yellow plastic ear tag flashing like a distress signal. I cruised closer. My heart stopped. The unborn calf's *back* legs extended from beneath her tail, its oversized hoofs turned skyward. *Breech birth!*

I raced back to the house and awakened my husband, Bill, and son, Scott. "We've got trouble!" I said. "Tag 402. She's got a breech! Hurry!"

Moments later, his sheepskin collar up around his ears, Bill climbed into the pickup beside me. "Don't worry," he said, "this cow has always raised fine calves. No problems." Scott, however, arrived armed with chains, an oxygen tank and vet supplies, dread preludes to what lay ahead for the unborn calf slowly smothering to death.

As we pulled up beside the cow, she stood pop-eyed with fear. "Good ole mama," Bill murmured, slipping a rope halter over her horns and under her chin before tying her securely to the bumper of the truck. "You're gonna be Okay."

Scott wrapped the glistening chains around the calf's cold ankles. "This is gonna be rough, Dad," he said. "Huge calf. Look at the heavy leg bones." He handed one chain to his father and gripped the other himself. "Ready?"

Bill nodded. "You pull first," he cautioned, "then me. Seesaw motion. Don't forget." Inch by inch, the hocks emerged, and finally the hips. Now the calf had to come out *fast* because if the chest wall expanded before its head was out, amniotic fluid could be sucked into its lungs.

Sitting on the ground, Bill and Scott each braced a boot against the cow's rump.

"Okay!" Bill said. "Pull!" Nothing. "Again!" Nothing. On the third try, the calf flopped out on top of them and lay motionless.

Scott scrambled to his feet. "It's a bull, Dad!" he shouted, unwinding the chains. "Geez! Look at the *size* of him! But he's not breathing!" He wrapped the slippery back ankles in burlap so he could get a good grip. Then, standing up, he dug his heels into the ground and swung the lifeless calf around and around like an Olympic hammer thrower. The centrifugal force, we hoped, would expel any fluid from the newborn's lungs. When he laid the calf back on the ground, Bill dropped to his knees and pumped its chest. Meanwhile, Scott clamped its mouth and one nostril shut, and blew into the other. The calf's rib cage rose and fell. We heard gurgling but no sign of life.

"But I can feel a heartbeat!" Bill insisted. "C'mon, fella...breathe." In desperation he poked a stalk of dry grass up the calf's nose. The calf sneezed and gulped for air. "Okay!" He sighed triumphantly. "Let's get him to the barn. Quick!"

Back at the barn, we laid the barely alive calf in a straw-filled stall beneath blankets and a heat lamp, and started an IV. Scott gave him shots of combiotic and banamine to reduce his swollen tongue, but his breathing remained shallow, his eyes unmoving. We'd been through this struggle before. We'd seen pneumonia strike. We'd mourned the deaths of beautiful, perfect calves that never woke up.

"Biggest calf we ever had," Bill said. "Probably 150 pounds. A real Goliath. What a herd sire he would have made!"

"But he's still alive," I said. "Maybe… just maybe … Do you want me to keep an eye on him the rest of the night?"

Bill nodded. I was glad he spared me the usual lecture, *Now don't get attached*, a habit of mine with the young, the sick, and the helpless. This time, however, the odds of survival were slim—no time for love, just hope, and a small prayer over and over. *Please don't die, little guy. We're going to need you one day.* I dozed.

Several hours later, dimly aware of the freezing winds whining through the rafters, I awoke to a sight I shall never forget. Curled up on top of the blankets beneath the heat lamp and right over the calf's heart, slept a wild, ugly, half-grown Manx cat. Because he had one ear torn off and a single fang jutting like a toothpick from his lower jaw, the children had named this untouchable creature Dracula. He glared at me through slitted eyes. Afraid to move, I glanced at the calf. My heart melted. Two enormous brown eyes embedded like chocolates in white velvet cushions stared into mine. *Goliath*, I thought, *you're going to make it!*

By morning I'd fed the calf two bottles of colostrum, "first milk" that we kept in the freezer for emergencies. Although he couldn't stand up, he drank greedily and searched my fingers for more. "C'mon, boy," I urged. "Get up!" But when he tried, he toppled and lay bawling and thrashing in the straw.

After breakfast, Bill and Scott made several futile

efforts to help him stand. "He's probably just sore from being pulled," Bill said, "but he's got to nurse. Let's splint him." They wrapped Goliath's back legs in cotton batting and slipped them into two sections of plastic irrigation pipe. Soon he wobbled around the corral like a clown on stilts and nursed relentlessly till his mother's udder hung like an empty glove.

An anxious smile crossed Bill's face. "That cow's never going to have enough milk to satisfy such a big calf. We're going to have to bottle-feed him." He looked at me. "Do you have time?"

Of course I did. In my mind Goliath was already a cuddly stuffed toy from my childhood. In my heart he'd become the storybook bull, Ferdinand, who sat under a cork tree all day smelling the flowers. That he'd have to be a good herd sire to pay his way was the farthest thing from my mind. All I cared about was that this soft, fluffy calf's huge, shining eyes brimmed with pleasure at the sight of me when I brought half-gallon bottles of milk. Also brimming with pleasure was one-fanged Dracula who licked away the sticky bubbles spilling down Goliath's chin and didn't stop until every hair on the calf's chest, knees and legs were scrubbed clean and white, and his hoofs shone lacquer-bright.

At first, Goliath accepted the cat's ministrations as though they were meant to be. But soon I noticed that every time Dracula left him in search of a hapless mouse, the calf bawled piteously. I wondered, was this the beginning of a strange and unusual friendship between a cat so wild no human hands could touch him but a bull with a heart could?

Several weeks later, when it was time to cut the splints from Goliath's legs, his weight was already up to 300 pounds. "Gosh, Dad," Scott marveled, "he's doubled his birth weight already! Do you think he'd win a first in the show ring?"

"Hey! Not so fast," Bill said. "He looks like he'll have the size, but don't forget, *it isn't blue ribbons* that make a good breeding bull."

Even as Goliath's extraordinary growth continued, Bill saw problems. Clearly our long-legged calf spent far too much time lying down. "Might help to put him out in a pasture," Bill suggested. "He can run with the other calves and get some exercise."

Not so. Goliath promptly chose the coolest place he could find, in the shade cast by our old adobe hut and the windmill creaking nearby. It wasn't exactly a fantasy cork tree, but he plopped down and seemed content. He didn't mind that his mama took off to graze all day without him, and instead of romping and butting heads as bull calves do, and terrorizing tarantulas and toads with stomping hooves, he seemed content to watch the other calves play. There, he waited for the feed wagon, rubbed his itchy horn buds against the windmill's cement foundation, and bobbed his head with pleasure at the sight of Dracula, who still found this quiet calf's soft, warm back a comforting place to be.

The children and I visited Goliath often. Unlike other calves that ran away if we came too close, this one didn't. His dark eyes shone expectantly. "He knows we're bringing Oreos," Jaymee said, skipping alongside me. Goliath loved Oreos, and the girls squealed with laughter when he licked the treat from their palms with his sandpaper tongue, chewed for several minutes, then probed each of his nostrils with the tip, savoring the lingering aroma.

At eight months Goliath weighed 800 pounds, by far the heaviest and largest of all our bull calves that year. But his back legs simply weren't right.

"I'm afraid we better steer (castrate) him," Bill said one night. "He'll make a great 4-H project for a kid."

"No, Daddy!" Becky cried out. "Then he'll be auctioned for meat!"

"He's too good to steer!" I said.

"Good for what?" Bill asked. "Odds are that a bull his size with a weakness in the hindquarters won't be able to mount a cow to breed her. Also, he might be sterile from the high temperature he ran at birth."

"We don't know that for sure," I argued. "He's only ten months old. Give him a little more time."

On weaning day, Goliath tipped the scales at one thousand and one pounds. Furthermore, I could lead my gentle Ferdinand around the barnyard without a ring in his nose, on just a simple rope halter.

Scott groaned. "You're making a real sissy out of that bull, Mom," he said. But I didn't care. Why didn't anyone understand that Goliath was just a little different from ordinary bulls?

Bill had other thoughts. "He'll weigh over 2000 in another year. That's over a ton of hay a month, plus feed, vaccinations, worming. Thousands of dollars! That's a huge expense for a bull that may or may not be able to breed."

"But we *need* a bull with his bloodlines," I countered.

"A new bull's expensive too! Isn't it worth the try?"

Bill looked thoughtful. "I'm going to start an artificial insemination program," he said. "It's the cheapest way to introduce a new bloodline, and safer than buying a new, unproven bull. Scott and I have already selected thirty of our top cows and heifers."

"That's a great idea!" I said. "Then you can give Goliath a chance. Put him out in the pasture a month after you've inseminated the last one. If any come back in heat, *he* can breed them, and when calves arrive a month late, we'll know they're his. It's a chance to prove himself."

Smoke curled from Bill's pipe. "We'll give it a try."

Winter closed in. Unlike every bull we'd ever had, Goliath's affection for other barnyard animals showed. He let chickens peck away lice from his ears and flies from his ankles. Rattleheaded doves shared his grain in the trough. Cactus wrens nested in his hay bin, and when he slept with his chin on the ground, Dracula crouched nearby, waiting to pounce on invisible mice stirred up by Goliath's nostrils puffing the dust.

The following spring, a month after Bill and Scott finished the artificial breeding, with doubt in his heart Bill turned Goliath loose in the windmill pasture with his select group of cows and heifers. Gone was my Ferdinand. The huge bull fixed his dark eyes on his future and snorted loudly. Lowering his rack of grappling horns, two thousand pounds of muscle and brawn pawed the ground, sashayed into the midst of the wide-eyed herd, and announced his intent in a deep, rich baritone.

However, like most inexperienced young bulls,

Goliath spent his entire first day making a nuisance out of himself chasing heifers that wanted nothing to do with him. A novice at the game, he didn't understand. Then, on the second day, a fetching two-year-old roused his passion to uncontrollable heights. Wheezing and sweating, he limped and lumbered after her, rolling up more square miles in a single afternoon than he'd covered in his entire life — and he wasn't alone. Boxing the foot-long tuft of hair on the end of Goliath's tail, Dracula clung and swung, making it doubly difficult for the bull to woo his beloved.

By nightfall Goliath's knees ballooned like footballs. His voice was a squeak, his chest cut and bleeding from vicious kicks bestowed by his intended. The following morning he lay in the shade of the old adobe, his chin on the ground. We carried hay, and water by the bucketful, to him. To our knowledge he didn't get up for a week, and all the while Dracula slept peacefully on his back.

During the months that followed the heifers bloomed, and our hopes for success with the artificial insemination program soared. It would be seven more months before we saw results — and eight before we'd know if Goliath had sired any calves.

Winter slipped into spring. The artificially inseminated calves began to arrive. With each birth, Bill and Scott tried to recall which of the two had handled the specific insemination. At the month's end, the score was seven to six. Scott won. But the calving ceased. "Damn," Bill said. "Only 13 out of 30!"

"How come so few?" I asked.

"We're just beginners," said Bill, shrugging his broad shoulders. "Like anything else, it takes practice."

"We should have put a *good* bull in the pasture with them, Dad," said Scott. "Now we've got 17 unbred cows."

"You're forgetting Goliath," I said, unwilling to give up hope.

Thirty days after the last artificially inseminated calf was born, Goliath's first passion produced twin bulls. Sixteen more healthy calves followed, all fathered by Goliath. Not one was too big at birth, and nearly all had the traits Bill had hoped for—long legs, stocky builds and calm dispositions. Still we couldn't help wondering. *With bad legs and problems of weight and mounting, how'd Goliath do it?*

And then I knew the answer. Goliath wasn't just the biggest bull we'd ever had, he was the smartest. Unlike other bulls, he'd discovered the chase wasn't necessary. *Patience* was the secret. All he had to do was wait in the shade behind the adobe, roll his upper lip back over his nostrils into a roguish grin, and when the right moment came the girls would come to him.

Over the years there came a gradual change to our ranch. Irrigation transformed parched lands into lush, green fields, and uniform herds of bigger-bodied, longer-legged cattle replaced the cows of the past. Only one thing remained the same. An old bull, surrounded by calves, rested in the shade of the old adobe near the windmill just as he had in his youth. Birds perched on his spectacular horns, singing their hearts out in spite of the cat with one fang atop his mahogany hide.

I wished time could go on forever, but, of course, it couldn't. Not for us, and especially not for Goliath. The afternoon came when my hand tightened on his last bucket of feed laced with two Oreos and his daily dose of "bute" to ease the pain of crippling arthritis. We had to put him down.

He didn't seem to know I was there till I knelt beside him to bestow the usual loving hug. When he opened his chocolate eyes the chill wind of a cold March night drifted up through the years. In my memory I saw a newborn calf struggling to live and once more I heard my wish— *We're going to need you one day.* Now—on this, his last day—I scratched the curly places behind his horns, right where he loved it most, and murmured farewell to the best bull we ever had. *My Ferdinand.*

We buried Goliath in our cattle graveyard beyond the White River Draw. There are many mounds there now. But from that first day and others after that, we often saw an old, wild Manx cat with one fang, curled up on top of the *biggest* mound—right where the heart should be. Dracula was mourning the biggest and best friend he ever had. We did too.

STARSTRUCK

Alex grew up to the twinkling of stars. Every night on the porch of their miner's shack that clung to a ragged peak near the top of Mule Mountain, he, his brother Zeke, and their Pa watched the changeless patterns in the blinking heavens. Sometimes they saw a star move. And one evening, later than usual, Pa reached up and caught one. He put it in his pocket.

"You didn't really catch one, did you, Pa?" Alex asked.

"You're dang right I did. Stars are made of gold."

Alex refused to believe him. They argued about it for years. Just Alex and Pa argued. Zeke never said a word.

Alex and Zeke were born under the sign of Gemini.

"Twins!" wheezed Estrella after her long walk from the adobe hut on the outside of Bisbee and her climb up the one hundred and thirty-two crude stone steps chiseled into the rocky Mule mountainside that led to the Timberlakes' crumbling shack. Too late for the first baby, fair, blue-eyed, unmoving, the midwife grabbed him by the ankles anyway and shoved him into the terrified miner's hands. "Smack him till he hollers," she ordered, and turned to the second baby, who shrieked his protest as he emerged into the chill, starlit night. "Now this'n's got a pair o' lungs that'd cause a cave-in!" she crowed. But Estrella's enthusiasm was brief. Ten minutes later, the dark-haired infant was still screaming, while the stricken father wept over his limp firstborn.

"He still ain't breathin," Ed Timberlake moaned.

The midwife grabbed the little red-bottomed, ghost-skinned creature out of his father's hands. "Yes, he is!" She pressed his tiny mouth to her ear. "He just don't make no sound."

Ed Timberlake grinned. "Seen and not heard, huh? Now that's gonna be a smart kid!"

Estrella nodded. "You should name him Zeke. Zeke means 'The Intelligent One.'"

"And what about th'other?" The miner jerked his head at the squalling twin.

Coins on the kitchen table winked at the midwife in the lantern light. Both babies had to be quiet, content, before she took her fee. Yet hadn't she, "granny-woman" of Cochise County, been born with the veil, proof of her gift of "second sight"?

A slow smile dimpled the corners of Estrella's mouth as she bathed the newborns, wrapped them in flannel rags, and placed them each in a separate crate Ed Timberlake produced from his tool shed. Although the shrieking baby's cries mounted, these sounds of new life plucked at her heartstrings like songs raining down from the heavens. She knew the ancestors were coming. She felt their presence seeping into her skin, their strength into her hair, and their wisdom whispering through her veins.

Closing her eyes, and swaying on wide, moccasin-clad feet, Estrella began chanting softly. She dipped her long fingers into the small beaded pouch concealed between the folds of her voluminous robe, and tossed a pinch of gold dust over the infant brothers. Then, driven by some unseen force, she lifted the squalling baby from his solitary crate and cuddled him firmly against his brother in the other. Small bodies touched. Two tiny hearts beat in rhythm, and the dark-haired infant wrapped his arm over his silent brother's chest and ceased crying.

Estrella scooped the coins from the table. "I can answer you now, Mr. Timberlake. You must name your second son Alexander... Alexander... 'The Protector.' For indeed he is."

On her way home through Brewery Gulch, the midwife heard a woman's cries, pulling her into yet another mud hut where she delivered a beautiful baby girl. No coins glittered on the kitchen table as she prepared to leave. *I don't need coins,* Estrella's heart whispered. "I have brought an angel into the world," she crooned. "Angelita... Angelita... 'Messenger of Love.'" Smiling at the young mother and her baby, she retreated into the night where a quick glimpse of the Bisbee sky revealed a brand new star in the heavens. It shone brightly between the Gemini — right where she knew it would be.

By the time the twins were four years old Zeke had mastered scissors. The first time Alex caught his brother with the forbidden tool, Zeke was seated on the dirt floor beneath the only window, a hole punched into the adobe wall to welcome the sun and the moon. It was nighttime. His golden hair shone in the dark like a halo, and stacked around him, piles of perfect five-point stars cut out of old newspapers shimmered in the moonlight. "How'd you do that?" Alex asked. Zeke smiled and tossed one into the air.

Stars are made of gold, Pa reminded them over and over again. *Pure gold.* Under Pa's guidance, Zeke had learned to draw stars easily. Soon he was tracing them on the dirt floor with his big toe or in the flour with his finger on the kitchen table when Ma rolled out dough. With a yellow crayon, he even drew them on their three kitchen chairs. Although Alex tried to draw stars too, his never looked like stars.

"Just a mess a' scribbles," Pa always said.

Earlier that night Ma had tucked the boys into bed. "Take care of your brother, Alex," she said, "while I tote the ironing to the mayor's wife." Although Alex fell asleep holding his brother tight so he wouldn't get away, he awakened suddenly to empty arms and the unmistakable snipping of blades and crinkling of paper. He scrambled out of bed. "Whatcha' doin', Zeke? You know you're not supposed to touch Ma's scissors."

Zeke beamed, picked out his favorite paper star and held it high against the moonlit window. The star seemed fringed in gold. "Wow!" Alex rubbed his eyes. "How'd you do that?" He knew how proud of his firstborn son Pa would be when he got home, and he pictured him nailing and gluing every single star to the walls, the ceiling, wherever he could find a space—except, of course, over the doorway where his precious single-action revolver hung loaded, ready. Then he'd open another bottle of whiskey and stagger back to St. Elmo's. "Gotta go tell the boys how smart my son is," he'd say. "Gotta go tell the boys," and Alex felt good inside, almost as if he had been the smart one and cut out all those stars himself.

The year the twins turned six, Zeke's skills with a

knife caused the villagers to stare, tongues to wag, and breaths to catch in wonder. Some saw magic in the miniature animals and birds the silent child now carved out of ironwood or stalks of dry agave. "But why all those stars?" they asked one another. "Why?"

Summer throbbed by, and rain took a swipe at the desert. From her adobe hut, Estrella smiled at the two small boys wading hand-in-hand through the shimmering mirages and up dry riverbeds in search of earth's treasures exposed by angry monsoons. The tall, fair-skinned child stuffed his pockets with chunks of jasper, agate, onyx and turquoise, while the smaller, dark-haired one sang like a mountain bluebird, and gathered sharp-edged river rock and flint, the tools his brother would need. Then, at the end of each day, she watched them crouching deep in the sand, sharing secrets only one could speak—yet both could understand.

When evenings came, a little girl with long black hair watched, too, from a distant place. Seated at the base of an ancient saguaro on the highest bank of a mighty arroyo, her small body cut a silhouette against the pinks, purples and reds of the Arizona sunset. Her name was Angelita, and she listened for the music.

The townspeople remarked to one another, "Have you seen them stars and little fetishes Ed and Maudie Timberlake's kid can chip out of rocks now?"

"Which kid?" newcomers would ask. "They got two, don't they?"

"Yeah. You know. The one that don't say nothin'."

"Oh... yeah!"

"Well, someday them little carvings is gonna be worth a deal a' money, more than we'll ever make pickin' a Bisbee mine."

"I hear ya, Joe, and you know somethin'? I think ol' Ed's already cashing in on them. Ever notice how he's always got one hand in his pocket as though he's scared to let something go?"

"Cash, I betcha."

"Or a gun."

Every time Alex overheard these fragments of conversation, a ping of unexplained sorrow crept into his heart and tears filled his eyes. He didn't know why.

Alex and Zeke were seven when the one-room school re-opened. The day the new teacher arrived from Salt Lake City by the stagecoach that pulled up in front of the Copper Queen Hotel, eager mothers pushed their children forward to greet her. Each child clutched a gift—a jar of prickly pear jam, a string of beef jerky or a freshly baked pie. Alex sang "America the Beautiful", and Zeke held up a star he'd fashioned from turquoise. He had tapped a hole through the center and threaded it with a fine leather thong so it could be worn as a necklace. "How lovely!" Miss Martin exclaimed. "Where did you buy it?"

Alex stepped forward and locked an arm around his brother's neck. "He didn't buy it, ma'am... he made it!"

The following day school opened, but a decision had been made. Zeke wasn't eligible to go to school. Instantly, Pa summoned Estrella and begged her to use some of her medicines and magic to "open his ears and loosen his tongue. Perhaps some gold-dust?" Estrella shook her head and smiled. "It is you who is deaf, Mr. Timberlake. If you can't hear him, you must listen to his brother's songs. Only then will you hear your first son speaking too."

If Zeke can't go to school, I'm not going either, Alex decided, but when morning came Pa dragged Alex by the ear down the steep stone steps and took him anyway. "You wan'na grow up stupid?" he said. Meanwhile, Ma tied Zeke to the woodstove so he wouldn't wander off looking for his brother, or get lost or hurt while she went to town with her basket to pick up the mayor's dirty laundry.

Alex struggled to escape Pa's grip when he shoved him through the schoolroom door. Tears threatened, and he balled his fists in frustration at Miss Martin. "Why can't Zeke come?" he cried. "He's lots smarter 'n me. Honest. Tell her, Pa! Tell her! Why won't you tell her how smart Zeke is?"

"You go along to work, Mr. Timberlake," the teacher said. "Your son and I have things to talk about."

"He ain't my son," the father snarled. "He's been a pain in the ass since the moment he was born. Nothin' but a big mouth and all noise."

"I'll have to ask you to leave, Mr. Timberlake."

The door slammed. Miss Martin sat down and took the weeping child's two small fists in her own hands, held them tight and pressed them to her lips. When Alex relaxed, she wrapped her arms around him, and for the first time in his life the little boy felt he had found someone to listen to his unfolding love story. "Won't be no bother, ma'am," Alex sobbed. "I promise. He'll be good... and he can sit in my chair with me—just like we do at home."

Miss Martin leaned closer now, and releasing her grip, she unconsciously toyed with the star swinging on the thong around her neck. She spoke gently. "Why doesn't Zeke sit in his own chair?"

Alex bit his lip. "He don't have none, ma'am. Pa's afraid he'll fall off and hit his head. Then he won't be smart no more." *And we won't have no money.* "So we sit in the same chair so I can hold him or catch him if he falls. Oh, please, Miss Martin. Please. I'll take care of him. I promise."

The teacher brushed dark wet curls from his face. "You bring your brother to school tomorrow, Alex," she said. "We'll see."

When Alex got home that afternoon and found Zeke huddled behind the woodstove at the end of a rope, his small face tear-tracked in soot, he vowed he would never leave his brother again. He'd take care of him forever.

Although most of the school children knew the Timberlake twins already, none could figure Zeke out, a child who towered over all of them, a strange boy who laughed and rocked and drew stars in chalk dust; a child who never spoke, nor would he cry no matter how much they tormented him. Furthermore, he didn't have to do numbers and spell words like they did. He was too dumb! That was it. Too dumb! Worse still, his brother was even dumber. All he wanted to do was fight. "Twins?" they'd jeer. "Why you don't even look like brothers!" That was all it took to keep Alex black-eyed and bloody-nosed.

However, things changed when Zeke began drawing

pictures of his schoolmates. A few pencil strokes and Tommy, Annabel, Angelita or Jimmy stared up from a piece of paper, a slate-board, or the sand on the desert playground. The children marveled, and it wasn't long before they realized that Zeke made a fine "Farmer in the Dell" too, one to be respected, perhaps envied because he always got to take the beautiful little Angelita for a wife, something they didn't dare to do for fear of their Pa's belts. After all, *who wants a wife, no matter how pretty, who spoke funny, didn't wear shoes, and brought only tortillas and beans to school in a rusty lunch kettle with a broken handle?*

Zeke seemed unaware of any of this. He loved Angelita. So did Alex. And when Miss Martin insisted Alex take a turn being the farmer, no one was surprised when he sang more wistfully than anyone, "The wife takes the child, The wife takes the child, Hi, ho, the merri-o, the wife takes the child." Then Angelita, although she barely understood a word of English, knew this was her signal. Stepping shyly toward the silent, gentle Zeke, she grabbed one big hand and pulled him into the center of the circle so he wouldn't have to stand alone.

"It's time you boys give me a hand in the mine," Pa said. Alex was ten when his world came undone. He had known for a long time that he'd have to quit school one day to help his father scratch for gold so there would be food on the table. But work in the mine? He talked with Zeke about it a lot because he knew his brother was afraid. "Don't worry, pal," he said. "I'll keep you with me down there. I only wish we could finish fourth grade. Don't you?"

Alex would never forget that first day underground. Pa climbed down the ladder first. It was dark in the mine, and Alex felt the uneven rhythm of Zeke's breathing as he guided him downward, one rung at a time. "I gotcha," he told him. "I gotcha. You're gonna be okay." Not till they reached the bottom, where Zeke clung to him in terror until Pa lit two lanterns, would he understand the extent of his brother's fear.

"Come on! Move it!" The familiar impatience in his father's voice made Alex's stomach grind, but he grasped Zeke's wrist and pulled him along the edge of a sharp jagged wall toward the "great room," a place, Pa told him, where the real stuff hides. They had only a few feet to go when Zeke stumbled and fell. He rolled onto his back and reached up groping wildly for his brother, but when Alex tried to help him up Zeke suddenly pulled away and stared at the ceiling above them. His eyes widened. His body relaxed. In the flickering lantern light, the ceiling of rock glittered with specks of pure gold, like stars in the heavens on a chill Bisbee night.

"C'mon." Ed Timberlake cuffed Alex on the back of his head. "There's no movin' him now. Leave him be. You and me have work to do."

Work? Where? Alex wondered. *What's wrong with right here?* There were veins of gold everywhere. He could see them himself!

"Follow me!"

Alex followed his father through an opening into a second chamber where dozens of niches pocked the rock walls like birdhouses. Most were carefully fitted with a frontal stone as though purposely designed to cover the contents stored behind. A stranger entering the cave-like room would never notice the dozens of movable surfaces jigsawed into rock. Pa removed the stone from one, reached inside and pulled out clusters of assorted stars and exquisite fetishes that Zeke had carved from jasper, malachite, obsidian, agate and turquoise.

"Some day we are going to be rich," he said. "People will come from all over the world to buy Zeke's work. Until then we must keep these hidden. No one must know they are here."

The years of work in the mine began, and Alex was sworn to secrecy. "You tell anyone about where these are hid and you're dead, boy. Understand?" His father's eyes glinted like white steel as they labored side by side, chipping niches into granite walls where the rock yielded little but an occasional thread of wire-gold, certainly not enough for Pa to have purchased those two draft horses Alex had heard the villagers gossip about, or the fancy wagon it was rumored he stored in the Contessa Camilla Vasquez' barn across the Mexican border. Yet all day long Ma lay ill in their small adobe home at the top of one hundred and thirty-two steps, barely fifty feet from the mine.

During the years that followed, Alex helped his father hack nitches into rock and remove tiny threads of gold, which they stored in tin cans, while Zeke sat in the adjoining tunnel and smiled at the stars glinting from boulders overhead. Seemingly content beneath his underground heaven, he pulled pebbles of turquoise and agate from his pockets and chipped emotions chained to his soul into magical works of art.

During those first few months, Alex couldn't decide which he missed most, school or Angelita. At least he saw her on Sundays at church, and afterward she came up to their place, arms laden with books and lessons prepared by Miss Martin. For a while, the three children spent time at the kitchen table laughing and playing games together, or they listened to Alex sing and strum the rusty five-stringed guitar he had found wrapped in a horse blanket and forgotten at an old desert campsite. Finally, he and Angelita studied, while Zeke sat in his mother's rocking chair carving, chiseling and whittling.

As the years went by, and Alex passed from one grade to the next under Angelita's tutoring, he often found himself marveling at the softness of her warm brown hand in his own calloused palm beneath the kitchen table.

One evening, just before he was to get his diploma, Alex couldn't look away from the copper highlights glinting in Angelita's long dark hair and the flecks of gold sparkling in her eyes in the dying rays of the setting sun. His mother had told him again and again, "It's not polite to stare," but right now, he couldn't help himself. He wanted to touch Angelita, run his fingertips down her face, hold her close.

"What's wrong?" she asked.

"You... you're so beautiful," he said.

They planned to meet at the mine one night when everyone was asleep. "Just the two of us," Alex whispered. "I want to show you Pa's secret." It seemed hours before Zeke's steady breathing assured him it was safe to slip out the door. In front of the mine, he waited for Angelita among the silvery shadows of a crescent moon.

Nocturnal creatures scuttled by and every crackle and crunch of tumbleweed or creatures of the night filled him with a longing that he could only express in song. At last she appeared. He put his arms around her. He wondered, was it her heart or his own pounding so hard against his ribs as he guided her down the fifteen-rung ladder? What did it matter? He loved her. He wanted to take care of her. "One day we'll get away from this town, Angel," he murmured, "the three of us. You, and me, and Zeke." Angelita understood. He didn't have to explain, and he loved her for accepting the way it was. "Yes," she murmured. "The three of us." He kissed her then with all the passion of his eighteen years.

Deep inside the earth Alex showed Angelita the treasure room and the small niches sealed with rocks that fit with jigsaw precision concealing the treasures hidden behind. "Zeke's masterpieces," he murmured. "Pa saves them all because one day Zeke is going to be as famous as Michelangelo and Leonardo da Vinci. His fetishes are going to be worth a lot of money."

Angelita gasped. "But your pa is already selling them, Alex. I've seen the money change hands." Angelita worked nights after school at St. Elmo's on Brewery Gulch Road. "Don't forget," she once told him, "your father boasts about the treasures his 'gifted son' has made, and how he has hidden them, 'like those Egyptian pharaohs hid their stuff, so good that no one can find 'em, except himself."

"Yes. But nobody knows where they are hidden... but us."

"And your father's friends at St. Elmo's," she reminded him. "He talks too much, and they talk too. Whiskey loosens the tongue. They say terrible things, mostly about his family on the other side. A beautiful contessa. El grande casa. Race horses and cattle. Mucho dinero. Are these lies? All of them?"

Alex shook his head. "I hope so."

But right now, Alex was alone with an angel he'd loved forever it seemed, so neither he nor Angelita heard the scuffing of deerskin moccasins in the sand at the top of the ladder, or later the tread of boots they knew so well, as they kissed again and again deep in the Timberlake mine.

It so happened, just before dawn on April 22, 1917, when Alex and Zeke Timberlake were twenty years old, a meteor shower struck the mining town of Bisbee, Arizona. Bright star fragments bombarded the surrounding desert, the mountains, and the wide arroyos of Cochise County. Windows shattered. Horses bolted. A cow was killed. Dogs and chickens fled in terror from the blazing rocks that plunged from the sky.

Alex and Angelita had planned to leave Bisbee that morning and get married. Zeke would go with them. The brothers were scrubbing up in the water trough outside the Timberlake gold mine when something hit Zeke in the head. Just before his brother was struck, Alex saw a trail of sparks arcing across the heavens. Then, a single, whining, blinding light hurtled toward them. "Look out!" he cried, throwing one arm over his eyes and reaching to protect Zeke with the other. Too late. He would never forget the hissing, sucking, squeezing sound, before his brother crumbled at his feet.

55

"Ma! Pa! Angel!"

Blood poured from Zeke's head as Alex dragged him over to their shack and laid him on the sagging porch. "Where are you, Pa? Get out here!" He cradled Zeke's head in his arms and rocked him back and forth. "Hurry, Pa! Zeke's dying!" he sobbed. "He's dying!" But instead of a steadying hand on his shoulder, Alex felt the touch of cold steel against the base of his skull followed by four clicks of a revolver's hammer being thumbed fully back, and the raspy rotation of the cylinder turning a cartridge into its deadly alignment with the barrel. "No! Pa!" he screamed. "I didn't do it! It was that goddam star!"

"Maudie!" Ed Timberlake barked. "Get the hell outa that bed and go find the sheriff!"

Ma, who had been too weak to leave the house for nearly a year, dragged herself down the rocky steps to the bottom of Mule Mountain for the last time.

The sheriff returned alone.

"Alex killed him!" moaned the crazed father. "The son-of-a-buck smashed my son's head in with a rock."

"Take it easy, Ed. They're both your sons. There must have been an accident here."

"It was murder! I know it was! He and that little tramp of his were takin' off, leavin' Zeke behind. The poor kid had even packed his bag."

"That's a lie!" said Alex. "You know I would never leave Zeke."

"Your boys are not kids any more, Ed. They're men. They're twenty years old!"

No sooner had the sheriff spoken than an icy wind-blast slithered up Brewery Gulch, sharpening Pa's stinging reply. "Twenty is plenty old enough to steal and kill, so while you're at it, book him on grand theft too."

Grand theft? Alex clenched his teeth. It had been ten years since the first time he caught Pa wrapping more than twenty of the exquisite fetishes in a red neckerchief and tying them to his belt. Alex's hand flew to his face as the staggering memory flashed to the surface. After all this time, he still felt the painful blow whenever that terrible moment flashed back as though it had happened only a day before. "Pa! Those belong to Zeke!" That's all he'd said. Ed Timberlake whirled around and struck him so hard on the cheek he heard the bone shatter. Now he knew for sure it was Pa who'd been stealing them all this time, but it hurt even more to realize that Pa had found the perfect opportunity to lay the blame on him.

Suddenly, Angelita dashed from the house, dropped to her knees and pulled Zeke's head onto her own lap. "Run, Alex!" she begged. "They'll hang you!"

Angelita was right. Too many had witnessed the fight between the brothers the night before. Pa had gotten a notion several months earlier to take Zeke with him to St. Elmo's Saloon. Zeke was well over six feet tall, and for reasons no one understood, unable to talk. Yet he was the darling of all the Bisbee housewives ever since his mother's long illness began. All day long, the silent young man chipped, carved, and polished jewel-like stars out of malachite, stars out of petrified rock and turquoise, and stars from exquisite mountain gemstones he and Alex collected for every generous lady in town in exchange for an apple pie, a dozen fresh eggs or a jug of homemade cider.

"Isn't that boy the sweetest soul ever?" they'd say.
"Too bad his brother is so... well, you know... "

"Ah, but have you heard him sing?" The younger
women exchanged glances. They knew better, and se-
cretly hummed favorite refrains from one or more of
Alex's songs, and sighed when his voice cast words of
love, sending forbidden thrills rushing through their wak-
ening adolescent hearts.

And the truth was... it was too bad that Zeke and
whiskey didn't mix. The miners, disenchanted with their
non-productive claims, made fun of Ed Timberlake's son,
"the one who don't speak, but drinks himself stupid every night on the stool at the end of the bar." It was even
more unfortunate that Pa wouldn't let Zeke bring his
knives and chisels and rocks with him, for there was nothing for those platter-sized hands to do but crack knuckles, twirl a glass, keep it filled, and pass useless hours
walking his fingers up and down the mahogany lip of the
bar. Nights were long, and whiskey led to taunting words
and ruthless fists. Pa often left his son lying in the street
after losing a fight and staggered home without him.

"Where's Zeke?" Alex would call from his bunk.

"Damned if I know," came Pa's reply, and Alex would
hurry down through the Gulch to bring his brother home.

And so it was, the night before stars fell from the
sky, that Alex arrived at St. Elmo's just in time to find
his brother, crazed with anger and bewilderment, holding a bar stool over the head of a miner already on his
knees. "Cut it out!" Alex yelled. Zeke spun too fast, strik-
ing his brother a crunching blow to the shoulder. Blind
with pain and rage, Alex caught Zeke with a roundhouse

to the chin, toppled him to the floor, and hollered, "You keep comin' down here and gettin' drunk like this—and I'll kill you!"

Everyone heard those words. Everyone saw the tears of confusion and shame coursing down Zeke's face. But not one soul would remember seeing Alex put his arm around his brother's massive shoulders to steady him for the steep walk home. Nor would they remember hearing him say, "Everything's gonna be okay, pal, just as soon as you and me and Angelita get away from this place."

After wrestling the revolver from his father and leaving the sheriff no time to draw, Alex forced the two at gunpoint into the tool shed and locked it. He and Angelita carried Zeke into the house and laid him on his bunk.

"I'll take care of him, Alex, until you come back... Hurry. You must leave... now!"

With his miner's knapsack strapped onto his shoulders, and his precious guitar gripped tightly in one hand, he hugged Angelita. "I'll be back," he promised, "when it's safe. You will have to let me know. I'll write so you'll know where I am." He heard Pa and the sheriff kicking and battering the shed door. "I love you," he murmured.

He vanished, traveling by night and sleeping in abandoned wolf dens by day. For the first time in his life, he was alone, completely alone. Alone without Zeke, he was nobody.

It had been easy to change his identity. The dark mustache was all he needed to pass for Mexican, and he had learned the language from his mother. Spanish was all she spoke. How easy it had been for him to be friends with Angelita when he started school. He'd been able to interpret for her because the moment his father came home it was "English only" at his place, and a slap on the side of the head should he forget. Memories of his mother cowering in the corner because she was "too stupid" to remember a word her husband might have taught her the previous day haunted him. Then there was Estrella, the midwife and village storyteller. Miss Martin invited her often to the school where she told the children stories of the stars in both Spanish and English so they all could understand. Alex had clung to her legends, and now as he traveled both night and day in boxcars through Texas, Louisiana and Georgia, he retold her stories to travelers who were hiding like himself, running away, or simply down on their luck. Unlike Estrella's storytelling however, he turned each one into a ballad or love song through the magic of his Spanish guitar. It wasn't long before the "Mexican cowboy" became known as El Estrelatto, the singer of stars.

Then came the war.

Angelita's first letter reached him through the army priest. *Zeke lives*, she wrote. *You mustn't worry. I will care for him, and the money continues to come through the good Father. Only a small scar remains on Zeke's temple, but the injury must have been deep. He no longer chips stones. Instead he goes to the mine each day where he sifts sand through his fingers and climbs the jagged hills , searching, always searching for something. When he comes home, he rocks in your Ma's old chair and stares at that picture of her holding the two of you when you were just little boys.*

Alex's eyes crowded with tears at her words. He had lost everything he loved. Ma. Zeke. Angelita. He yearned to go home.

Another letter reached him in Germany. *Although the charge against you has been changed to attempted murder and grand theft, WANTED posters for your arrest are still posted all over Arizona. The checks from Alejandro Estrelatto are never questioned, but oh I miss you, my darling.* Months turned into years. The war ended.

Alex had no home to come home to. Post office boxes, month to month, kept him in touch with Angelita. *WANTED signs still smear drug store walls and telephone poles,* she wrote, *and posters continue to read REWARD.*

Despite a shattered leg, Alex worked on railroads and cattle drives during the decades that followed. More and more frequently, however, he was invited to sing in local bars. Word spread about the talented, handsome singer and his ballads that made both men and women cry. County fairs begged him to come. He pulled in crowds. Then came invitations from churches. His music brought young people back to Jesus, back to the Lord, back to God. Families flocked to wherever the singer was. They insisted it was God they came to worship. Most knew better.

El Estrelatto sang songs that made teenagers sigh and old people cry. They asked for their favorites, over and over again. "Mensajero d'Amor" (Messenger of Love), "Estrellas Enchan de Oro", (Stars Are Made of Gold), and "Me Para Solo", (I Stand Alone). El Estrelatto became a national idol. In the 1930s it was radio, in the 1950s television. When, "Hermano Que No Habla", (A

Brother without Words) was released, the world went crazy, and Angelita's last letter found him in Las Vegas. *Please come home,* she wrote. *We need you so.*

Zeke? Something had happened to Zeke. He was sure of it.

Night closed in on Alex's Chevy pickup as it crept past the elevation sign. Five thousand nine hundred and eighty-seven feet. Straight ahead loomed Mule Pass, the tunnel, built since he'd left, that linked the Arizona mining town of his youth to the world beyond copper, silver and the lust for gold. He released his grip from the neck of the guitar resting on the seat beside him, lifted his hat from his brow and ran his fingers through his steel-gray hair, before brushing his thumb over the envelope sticking out of his shirt pocket. *Angelita... Angelita. I have never stopped loving you,* he sang, as his pickup dove into the tunnel.

Later that night, down at St. Elmo's and in the restaurant at the Copper Queen Hotel, townspeople and tourists claimed that, just before the snow began to fall, they saw a stranger in a pearl-bellied Stetson limp slowly up the one hundred and thirty-two steps to the wind-battered shack so long neglected at the top of the mountain. Why would anyone want to visit the old Mexican lady who lived up there and took care of some old crazy with a rock in his head? Why?

They had no way of knowing that the higher the stranger climbed, the more enchanted his world became. When he reached the top, early flakes of snow traced a lacy carpet on the ground, and in the shimmering glow of lantern light shining from the window, ice crystals sparkled on brittle branches of yucca and mesquite, lending a touch of magic to the place Alex had once called home. Tumbling back into the world of his childhood, he tried to wipe fog from a windowpane where once yawned an open hole in the adobe wall. Suddenly, the window was no longer there and the image of a little boy reached out to him. He had scissors in his hands, and his hair shone like a halo in the winter moonlight. He wasn't doing anything really wrong, just sitting on the floor smiling and cutting out stars. Alex fought the lump rising in his throat. His hand reached out for the pair of scissors, but his fingertips touched the cold, cold glass, and the image of the brother he loved faded away. He blew away another dusting of snow on the sill and peered again inside.

A beautiful woman with silver racing through her long black hair sat at the kitchen table, sewing. *Angelita!* Beside her, his hair white as the snow on the highest peak of Mount Lemmon, sat an old man with a picture frame pressed to his chest. Alex looked closer, and through a painful blur of tears he saw a five-pointed scar punctured into his temple. *Zeke!* He could stand it no longer. He threw open the door. Angelita was in his arms.

Tears tracking the wrinkles that landscaped his face, Zeke rose from his chair and wrapped them both in his long thin arms, the brother he thought he'd lost forever and the angel who had cared for him for so many years. Finally, he let them go and shuffled slowly across the room to the peg where Pa's jacket still hung.

"Has it never been worn?" Alex asked with a smile.

"Oh my, yes," Angelita said. "Zeke slipped it on every day before leaving for the mine. I used to watch him through the window, crawling on his hands and knees where the water trough used to be. There were times he would rock back and forth, stretch his hands to the sky, plunge them into the earth, scratch with his nails till they bled, then open his mouth like a wild thing screaming in the jaws of a trap. But no sound ever came. Over and over he sifted sand through his fingers. Searching, always searching. Some days he climbed among the rocks on the higher slopes and other days descended one rung further down the ladder into the bottom of the mine. But whatever the weather, he never left the house without wearing that jacket... until about three months ago." She put her hand to her throat as though she felt pain there. "He came in one day like always, hung the jacket back on the peg, sat down in his rocking chair and wept. He hasn't left the house since."

"... and that's why you wrote 'we need you so.'"

She nodded. "He's stopped eating, Alex. He wants to die. All that searching kept him alive, till one day he'd had enough. I knew the only thing that could save him was you."

They watched Zeke run his arms into the tattered sleeves of the ragged old garment and pull it up over his too-thin shoulders. His smile was that of a small boy as he shoved one hand deep into the jacket pocket and shuffled toward them. He pulled his hand out, and on his palm rested a golden nugget, the shape of a perfect, five-pointed star.

It is said that long before dawn the following morning word had spread among the early risers of Bisbee that El Estrelatto was in town. An eager voice at St. Elmo's confirmed the news. "I saw his truck coming through the Gulch last night, just before the snow started to fall."

"So did I," another replied. "A black pickup... stars painted all over it."

"And his name was written in gold."

"What name?"

"El Estrelatto!"

"El Estrelatto?" They gasped.

"Why is he coming to Bisbee?"

"Will he give a concert?"

Finally, the most reliable witness of all, the bartender who was last to leave St. Elmos's the night before announced, "With my own eyes, I saw his truck parked at the bottom of the steps leading up to the old Timberlake place."

"You mean that old mud brick adobe near the top of Mule Mountain where they used to say that Mexican lady quilts and keeps house for some old feeble-minded soul?"

"That's the one."

"Why would El Estrelatto park there?"

"Who knows?"

"Did he go up the steps?"

"Could it be that he... ?"

Eyes locked in dim memories, and from further down the bar an old-time miner said, "Yeah! The story goes that the ol' crazy's brother tried to kill him because he was after his girl. I was about forty at the time, but I remember the meteor shower, and the wild tale the Timberlake kid made up about one of them stars hitting his brother on the head. Damn near killed him. He was unconscious for years. Anyway, the brother managed to get hold of his Pa's gun, took all his gold, and locked him and the sheriff in the tool shed and took off. He's never been back since."

"Whoa!"

"Yep! That's the way I heard it. He got away with attempted murder and grand theft and skipped town."

"Wow!"

By late afternoon, crowds had gathered in the Gulch. It was Saturday. Rumors spread. "El Estrelatto is in town." There would be a concert in Bisbee that night like they'd never heard before. Maybe, if they hurried, they could get autographs.

"He's staying up at the old Timberlake place."

"El Estrelatto!" The balladeer's name was on everyone's lips. Fans rushed to the spot where his truck had parked the night before. They pushed, they pulled, and they shoved one another up the one hundred and thirty-two stone steps. Some slipped and fell but all pressed on, eager to be the first to touch him, to meet him in person, to hear him sing. Oh, the music! Eyes round, boots crunching the frozen ground, voices escalating, they persevered. Shouts became symphonies rising like frost balloons in the chill night air—until they reached the top.

"Don't look like no one's lived here in years," someone whispered. Indeed, the door of the ancient adobe hung open. A broken stovepipe and a few rusted wires shifted back and forth in the wind over the corrugated metal roof. Shards of glass glinted like ice chips on the sill of an open hole in the mud-brick wall, tempting a few to peek inside at three rotting chairs laced with cobwebs and a daguerreotype photo of a mother and her two little boys, cracked and broken on a dust-coated kitchen table.

"El Estrelatto?" Voices persisted, cracking and pleading in the night. "Where are you, El Estrelatto?"

Only a handful noticed the wide moccasin footprints in the snow beneath the hole in the sagging wall. Fewer still paid any attention to the caved-in mine shaft barely fifty feet from the shack. Nobody cared... except later when the grannies, the spinners of tales, would recall a great deal more about Bisbee of long ago, the rumors that became facts over tongues and times. It's these old folks who tell stories of a village midwife who cast spells on old men and little boys. They speak of a miner who made millions yet never mined the gold in his claim. Some recall a tale of two little brothers—one who could not speak but possessed a strange and wondrous talent, and the other sang like an angel.

There are also those who can sew life into pictures, and the village sewing guild takes pride today in taking these stories one step further with needle and thread on six-inch squares and stitching history into quilts. Hanging in the museum and abandoned store windows along Brewery Gulch Road, patchwork masterpieces of small children playing "Farmer in the Dell", and rocks on fire falling from the sky, catch the eyes of passing tourists. They stop to cup their hands around their faces and press their noses to the glass to better see the fine needlework on cotton squares where a little boy strums a five-stringed guitar, a barefoot angel walks hand-in-hand with two little friends, and a golden-haired child sits in the moonlight on a hard dirt floor clutching a star in his hands. The legends are there, the embroidered stories of lives lived long ago, forever preserved in time by needle and thread—and rags.

Men have memories too. A few shots of whiskey down at St. Elmo's is all it takes to restore sound and scenes of the meteor shower that struck their little town nearly one hundred years ago. Many of the tales have suffered from hallucination and become twisted, exaggerated stories of an astronomical phenomenon that struck Bisbee on April 22, 1917. "Terrible thing happened. A young man was struck by a falling star—or did his brother try to kill him with a rock?"

Who knows what's true and what's fantasy? Life stories are the stuff legends are made of, but what the storytellers of today like to tell about most is that one starlit night in Bisbee, Arizona, a famous balladeer came to town. Was he a miner's son from Bisbee? Did he try to kill his brother? Nobody knows for sure. But it is said that his name was El Estrelatto, a singer of stars. He brought love songs played on a five-stringed guitar to the world, and one night he came to Bisbee. His fans rushed to the top of Mule Mountain just to hear him sing. Those afraid to dream – or get caught up in the magic of the moment – claim he wasn't there. They walked away.

But the believers stayed, drawn by the music that came from the sky... ... songs that fell from Gemini.

Later that night in her adobe shack that rested between the ranch lands and the new road on the outskirts of Bisbee, there still lived an old woman, the only person in the county who knew for sure what happened. She had been drawn to the doorway in her moccasin-clad feet by the sound of music showering down from the skies. She saw a pickup truck drive by. A black pickup—gold stars glittering on its shiny sides. At the wheel a man wearing a pearl-bellied Stetson was singing. He waved to her. Against the opposite door sat another man, his white hair a halo in the frosty night. A beautiful woman was squeezed between them. Estrella hoped they wouldn't crush her wings.

The old midwife smiled beneath the glow from Gemini. "Everything is just as it should be," she murmured. "Just as it should be."

A LOVE THAT KEPT HIM WARM

One chill December morning, my husband, Bill, and I spotted a tiny hummingbird warming his tummy feathers in front of a spotlight attached to our cattle ranch's desert-based windmill. "Poor little thing," I murmured, "It looks like he forgot to fly south."

"Forgot?" Bill frowned. "Birds don't forget."

"Then he's hurt?" I said. *Surely he will die.*

Hummingbirds always left our Arizona cattle ranch in mid-September to migrate more than 2000 miles into the tropical jungles of Central America. There they would be safe from the freezing winds that swept up from the Gulf of Mexico into the desert lands of Cochise County in the great Southwest. But why, this year, did one tiny, jewel-like bird stay behind? How could he possibly survive?

This was a Costa hummingbird, among the smallest of the species. Although he would weigh less than two-tenths of an ounce on a scale, he had to eat insects and nectar every minute to replace the calories he burned. But this was winter, and now he was alone in a corner of the world where temperatures could fluctuate 65 to 75 degrees in less than 24 hours. Nothing was left to keep him alive, and nothing to keep him warm—except the shelter he sought in the nearby adobe hut and—a light bulb.

On this particular morning, the mesquite, manzanita, and prairie grasses crackled and snapped in the frigid air. We did the only thing we could. We hung a feeder of sugar water close to the warm spotlight to keep the liquid from freezing. As we wired the glass container to the windmill, the little bird buzzed bee-like around our heads. "He seems to know we're trying to help," I said. *But would sugar water alone be enough to sustain him?*

Perhaps it was, because when spring came our hummingbird was not only still alive, but his colors seemed brighter as he darted among banks of budding wild roses and zoomed across fields of lavender alfalfa. We liked to think his second chance at life was due to our helping hands. But over the next twelve years we learned his survival was part of something far greater—the most powerful link to life in the Master's plan—love.

As days grew warmer, we watched this busy, living gem sip nectar from flowers, catch gnats in midair with his tongue, and skewer moths and wasps with his needlenosed bill, and it wasn't long before we realized that most of his daylight hours were spent defending his territory against the return of the enemy. Like a miniature gunship, he bent his tail in rudderlike fashion and flew backward and forward and upside down. Chattering angrily, he attacked woodpeckers and doves, and chased away birds *ten times* his size. He fought relentlessly, even against his own kind, for what he seemed to feel was his.

"Wow! He's a humdinger," Bill said one day. "That'd be a good name for a little creature who's always at war with the world."

I agreed. But there had to be a reason for battle, a purpose to his territorial behavior. Only when evenings came did the whirring wings gave way to quieter moments, and we'd see the little bird perched on the rotating windmill rudder. Facing the south, he waited.

Love arrived in April—a drab little hummingbird in shades of dusty gray. From his lofty roost, Humdinger watched this tiny shadow of his own radiance build her walnut-sized nest on a forgotten lasso looped over a nearby fence. He forgot about war. Instead, he took up sky dancing, tracing intricate patterns against fiery sunsets, as love crept into his heart.

Soon, the tiny female joined him and a mating ritual followed. Scolding and squeaking in morning's splendor, they showered in misty rainbows cast by pasture sprinklers. They preened and puffed amid dawn-lit spider webs and played hide-and-seek among snowy yucca bells. At last, exhibiting his feathers in their most vibrant, jewel-like colors—ruby, emerald and amethyst— Humdinger wooed and won his tiny Cinderella in a graceful nuptial dive.

Two white, pea-size eggs appeared in the nest. Two tiny babies hatched and were fed and nurtured by two parents who loved them. Not until they were mature enough to tackle life on their own did they finally fly

away. After that, it wasn't unusual to see the evening silhouettes of Humdinger and Cinderella, two of earth's tiniest creatures, perched side by side on the tail of the windmill rudder. Then, September came. And Cinderella flew away.

Why didn't he go with her? I wondered anew. Unable to accept that I was witnessing one of nature's great mysteries, I continued to think something was wrong with him, or that Cinderella had simply tired of him. Could he survive another winter alone?

I needn't have worried. Humdinger did survive, and year after year when spring returned, Cinderella brought love to his life again. Wars were forgotten in lieu of sky dancing, taking showers, making love and raising babies. And every autumn, when the September nights dipped and winter snarled down from the mountains, she left him.

For twelve years, Humdinger never left the ranch. We replaced the light bulb each November. In March the wars began. And since wildlife patterns and habits are changeless, twilight found our little hummingbird sitting alone on the windmill rudder, facing south—waiting— for Cinderella to return.

His thirteenth winter came. *Will he still be there?* I wondered as I filled the hummingbird feeder with sugar water and took it with me, for I am a creature of hope. There was a cold snap in the air that night. Bill turned up his sheepskin-lined collar. "Maybe I'll go with you," he said. Heading toward the pasture, the walk seemed farther than it used to. I thought about other years, when the tiny bird would see us coming, his brilliant jewel-flashes glinting in the night with colors that appear only in dreams. But this night seemed too quiet. I braced myself for disappointment, moved closer, and hung the feeder at the bottom of the windmill. A calf bawled in the darkness, and an owl hooted its warning. No Humdinger. We turned to head home.

That's when I heard a whirring of wings, and there, warming his tummy feathers against the spotlight, hovered a shabby little hummingbird. Gone were the flashing colors of youth, but he was surviving, waiting for love, just like we all do. Then, when April comes, I mused, with luck he'll still be around to remind us... it takes more than a lightbulb ... *It's love that keeps us warm.*

"YO SOY TOMÁS"

"My name is Tomás"

I had just dropped my children off at the rural school bus stop when a sudden blast of wind tossed my pickup like an unwanted toy. I whirled a gloved hand in circles against the frost-covered windshield. Ringed in white, Arizona's cactus and desert shrubs cringed in the grips of winter.

Suddenly, another gust revealed a flash of red behind a mesquite clump. I held my breath and brushed hard at the persistent fog blurring my view. There it was again — a red wool cap. And it moved!

Who could be out in this deadly cold? I wondered. Our ranch huddled at 4500 feet near the Mule Mountains of Bisbee, 100 miles east of Tucson, 25 miles from the Mexican border and 15 miles from the Douglas prison. *The prison? A red cap?* Had someone escaped? Had I missed the radio alert?

I coasted to a halt but left the engine running. "Hello!" I called, opening the door cautiously on the passenger side. "Who's out there?"

"Buenos dias, señora." Bushes trembled, and a living scarecrow emerged, a man no taller than five feet six inches, his eyes... dark pools of hope.

"Buenos dias," I said, horrified that the tattered jacket he wore offered little warmth to his stooped, emaciated body. Shoes, long gone, had been replaced with rotting burlap tied to his feet with baling twine. He struggled to blow warmth into ungloved hands and seemed reluctant to step any closer.

"Hase frio!" *"It's cold,"* he murmured. Starving, sunken cheeked, stiff with pain—this man smiled.

"Tengo hambre, senora." He placed a hand on his stomach. How long had it been since his last meal? I wondered. How many miles had he walked in hopes of a job and money for his family in Mexico who hungered too?

I opened the door wider. "Vamos, mi amigo," I said. "Café es caliente a mi casa."

"Graçias, Señora." He jumped inside. "Yo soy Tomás."

Tomás and I drove in silence down the last few miles to the ranch headquarters and the warmth from our wood-burning stove, where the lingering aroma of bacon, eggs and home fries added to the welcome.

Bill, my husband, pipe clenched in his teeth, and our 20-year-old son, Scott, were still at the table hunched over coffee mugs trying to thaw out after morning chores.

"I brought someone with me," I called, urging Tomás to come in. But the little man stood on the doorstep, twisting the battered red cap in his hands.

Scott came to my rescue. After several minutes of fluent Spanish, Tomás sat down in a chair close to the stove. Eyes downcast, he gripped the tiny wooden cross hanging from a leather thong around his neck and murmured a prayer. He ate cautiously, a sign of previous experiences with the perils of wolfing a meal after starvation, while Scott shared his story with us.

Tomás was an illegal alien from Mexico. He had walked over two hundred miles existing on tortilla remnants. After crossing the border he traveled only at night for fear the United States Border Patrol might catch him. By day he slept buried in the sandy banks of arroyos. His family: a pregnant wife, three children, a blind uncle and a crippled grandmother were counting on him for survival. They had no money for food. There were no jobs. He had to find work.

Tomás was a "wetback," a popular label for Mexico's poorest since the turn of the century. Many, desperate for work, swam across the Rio Grande dividing their country and Texas and emerged on the U.S. shore, their backs glistening with water. To the Mexican Policía and the United States Border Patrol they are "illegal aliens." To the farmers and ranchers of border states they are "workers" they could not manage without. Tomás was our first encounter.

We knew it was against the law to hire this man. We shuddered at the thought of the hefty fine should we be caught. It was our duty to notify Customs at once should one show up so they could return him to the "other side." "But everyone else hires them," we reasoned; the chile farmers, pecan-orchard and cotton growers, as well as the ranchers. Why shouldn't we?

Ranch help was impossible to find. Although Bill had spoken with the Department of Economic Security many times, the answer was always the same. "No one wants to work when they can collect unemployment and do nothing." We had often offered jobs to the "drifter," but a remote ranch is a lonely place for a young man, and isolated rural bars offer the only recreation for Saturday nights. After Saturday payday, we rarely saw one again.

It was nearly February. Calving season had begun. Fields needed to be prepared for planting. The reality was, we needed Tomás as much as he needed us, and it took no time at all for him to clean the spiders and scorpions out of the old adobe shack and get the fire going in the potbellied stove. A cot, blankets, canned goods, and a 200-foot extension cord so a portable radio that had belonged to one of our older children could bring a little music into his life, and the smile on his face was brighter than the stars in the western sky. "Es bueno!" he said happily.

For the next fifteen years, Tomás accomplished the work of four men. He cleaned stalls and corrals, helped round up cattle, saddled horses, painted buildings, repaired tin roofs, raked, swept, and cleared dead mesquite from the land. And he accepted the twelve- to fourteen-hour days with a beaming smile.

Our younger children, Becky and Jaymee, ages six and three, tagged after him like puppies. He chattered in Spanish, they in English, and everyone understood. When Tomás sang, they sang. When he taught them how to whistle, Bill and I almost went crazy.

The sending of Tomás's hard-earned salary to his wife in Mexico remained a mystery. He could neither read nor write, but in his torn plastic wallet he kept an address written in the exquisite penmanship of a priest. To this address we sent a weekly money order made out to his wife. Her name could not appear on the envelope, but when letters came addressed to our box, Tomás assured us she had received it. He always kept ten dollars for himself.

Besides the ten dollars, Tomás kept a black-and-white snapshot of his family taken in front of the mud-brick house he had built in Mexico. Seated on the ground were his wife, two-year-old son, his grandmother and a blind uncle. In the doorway, framed against the bare white wall, stood a little dark-haired girl no more than three, with a baby cradled in her arms. "Mi nuevé nino, (my new son)" Tomás said proudly. Then pointing to a little gray bush beside the open front door he added, "Mi rosas." His rosebush.

That first year, homesickness struck after six months. Tomás missed his family, and there was another new baby girl he had never seen. His eyes misted when he told me he needed to go home for a while. By now he looked twenty years younger and thirty pounds heavier. With the $250 dollars he had saved out of his weekly income, he asked me to buy presents for his family; a wristwatch for his wife, toy trucks for his boys, dolls for his little girls, cigarettes for his blind uncle and a bottle of tequila for his grandmother.

That night when I dropped him off at a safe place near the border, he looked like a little round Santa in his new red knit cap and flannel shirt, with his knapsack on his back. Bill gave him extra money for bus fare from Agua Prieta, but years later we found out that he always saved that money for food and walked home.

After Tomás left, we worried. How would we manage without him? American labor remained impossible to find. Green cards for workers took years to get. But early one morning, a few weeks later, Roberto emerged from the hay barn. He was gray and old, he shook, he couldn't see very well, and he was hungry. "Amigo de Tomás," he said brightly, and he needed just a few weeks of work. When Roberto left, Ricardo appeared, and after him, Manuel. Soon a pattern was set. Tomás, afraid of horses himself, sent Vicenté to break them. The year the tractor engine needed rebuilding, Carlos came, a "bueno mecánico." Jesús, Librato, Ramon and Javier, all were Mexican farmers, cowboys, carpenters or good at everything. *Never* did one steal. *Never* did one complain. Each shared a dream.

At 5:30 one morning, I opened the door to clouds of dust, *three* "amigos de Tomás" raking the barnyard. Horse stalls cleaned, egg baskets filled and the chicken coop spotless, these men were hungry. Feeding three more mouths was no problem, but jobs for three? Where? I telephoned ranching neighbors. "Yes. I could use some help," said one. "I've got several miles of fence to build. Can you bring them halfway? The wife will meet you in a blue van."

So I piled the children into the front of the pickup while the men lay down in the truck bed and covered themselves with a canvas tarp in case we encountered the Border Patrol. We headed down the twenty-two-mile stretch of dirt road to Tombstone, ever alert for the inevitable heart stopper, the paddy wagon coming in the opposite direction. Only one appeared. Fortunately, he was in a hurry. He waved, grinned, and shot by, and I thanked God he didn't have time for the usual visit, for such is the way of life on open country roads.

I met "the wife" near Bisbee. We exchanged recipes, the children's 4-H and FFA activities, calving tales, and finally, the "mohados." I told her about our faithful Tomás.

"Aren't you lucky?" she said. "We had one like that for years too."

In time, we became increasingly aware that our Arizona ranch, and others like it, lay like a pot of gold at the end of a rainbow along the borders of the southwestern desert lands. The "gold" was hope for thousands yet to cross over who could find no work in their own country and were willing to take on the ultimate sacrifice, a life of loneliness, in exchange for the comfort and survival of those they loved.

For fifteen years Tomás returned in January, delightfully critical regarding the work done by his amigos in his absence, and proceeded to whip the ranch back into shape. He knew our schedule and needs, and when he left, "temporary" help from Mexico continued to appear when he knew we would need it most.

From these hardworking men we learned firsthand about real poverty. Shoes, if they had any, were wired or twined to their feet. Holes in soles were caulked with plastic bags and scraps of cardboard. They always needed clothes. Socks were such a luxury they brought tears of joy. As for boots, Bill's size 10 1/2 EE magically fit them all.

Tomás was an expert at everything. One day, he gazed at our tractor with such longing that Bill asked him if he could drive one. "Si!" he replied, dark eyes sparkling. But we worried. What if the Border Patrol spied him through high-powered binoculars from Land Rovers, helicopters, and airplanes? Harboring illegal aliens promised a huge fine, but here was Tomás, eager and ready to drive the tractor, and Bill still had two hundred acres of alfalfa to plant. From then on, it wasn't unusual to see a cloud of dust in the wake of our green John Deere plowing distant fields, the "hired hand" at the wheel wearing one of Bill's cowboy hats and a red neckerchief over his face—protection, of course, from the dust. His name was Tomás.

More than anything Tomás loved my chickens. I had 200 Araucanas hens, a breed that lays green, blue, buff and turquoise eggs. Enchanted with these chickens, he spent hours singing to them while keeping the coop spotless. "Música" brings more eggs, he insisted. From then on, I played classical tapes in the hen house each time he left. Egg production doubled.

Sundays were melancholy for Tomás. After morning chores he spent the afternoon doing his laundry and hanging it out to dry on a barbed-wire fence. Then, stretched out in the dappled shade beneath our cottonwood tree, he listened to Mexican music on a portable radio and gazed at the latest black-and-white photo of his family, ever alert for planes overhead or paddy wagons on distant roads.

As the months became years, Tomás seemed to stand taller than when he first entered our lives on that bleak, cold morning so long ago. One day he showed me the latest photo of his family. There was always a new baby. "So many bueno ninos," I said. To my surprise his smile dimmed. His lips trembled, and in that fleeting second I saw the quiet, inner strength of this little man we had grown to care so deeply for. I wondered how much longer he would stay.

The year came when we sensed his time with us was coming to an end. Tomás often talked about the garden he had planted by his house. His family now grew all its own vegetables and some fruit. The children worked hard, and this year when he returned home he planned to buy a burro and pack apples, oranges, squash and corn in baskets on its back to sell in nearby villages. He was going into business for himself—and Maria, his oldest daughter, was getting married.

When June came, we felt a heavy sadness as we bade Tomás good-bye at the border. This time it was daylight, and he was dressed in jeans, a cowboy hat and a new plaid shirt. He would blend easily with the ordinary tourists and Mexican Americans heading "across the line." No one would question him, or even ask to see his ID. No one would know or even care that this middle-aged Mexican man had spent nearly half his life in hiding—or that tucked in his backpack was Maria's first pair of shoes just in time for her wedding.

Bill handed him a bus ticket and extra money for the bride and groom. Tomás beamed. "Gracias, señor," he said, over and over again as he reached for his own new wallet to put the money inside. That's when he hesitated and pulled out a photograph we had never seen. It was in color!

"Esta toda mi familia," he said proudly, holding up the picture for all of us to see. His whole family! His wife, his grandmother, his blind uncle, and at least fourteen smiling, barefoot children were sitting on the ground. These were the loved ones for whom Tomás had toiled, so far away, and for so many years.

The same little house stood proudly behind the smiles, but now, clinging to the roof and adobe walls, a giant rosebush bloomed. A little girl stood in the doorway, her dark hair and eyes framed against the white wall. She held a baby in her arms, just like her big sister had in a photo of long ago. But, this time, the wall was no longer bare. A cross of red roses had been hammered there. "For you, señora," he said, and pressed the picture in my hand.

I wanted to thank him, but my throat ached. All I could do was press the gift to my heart because in that one shining moment I knew this gentle man had handed me the key to his life. In a simple photo he had captured his dream, the same dream we all have when we believe in the strength and power of LOVE.

I took a deep breath and hugged him. "Adios, mi amigo," I said. "Vaya con Dios." And he was gone.

As we turned to drive home, a soul-stirring memory rushed up through the years; a winter morning... a flash of red... and eyes, dark pools of hope. It was then I heard that gentle voice, "Yo soy Tomás." And the tears came.

THE TRUTH ABOUT CALEB

It's now or never. Caleb eyed himself in the side-view mirror of his ancient pickup and closed the door of our old adobe shack where he was living for a while "between jobs." His sweat-stained hat did little to cover the palomino-white hair grazing his frayed collar, but what difference did that make? He was down to his last stick of beef jerky and a can of beans. It was time to go to Belva's. He'd put it off long enough.

Tucked among the yucca and the prickly pear near the crossroads of Earhart, Arizona, population 67, Belva's Country Store and Café baked in the desert sun. Nevertheless, the scattered ranchers and farmers, and families who dwelled in the surrounding crumbling adobes and battered trailers, depended on Belva. She stocked diapers and tractor parts, canned goods and udder balm, things folks need who live 100 miles from town.

When Caleb entered the store, a large woman in her middle sixties emerged from the rear. Her hair was cotton-

candy-blue and her skin a creamy cameo. She wore prism-thick glasses, and the biggest dog he'd ever seen leaned against her protectively.

"Howdy, ma'am." His heart swelled with a feeling he had long thought dead. "I'm caretaking the cotton gin on Frontier Road. Name's Caleb."

"Of course it is, and it's about time you got your fanny over here, Caleb," she boomed. "I'm Belva."

"Pleased to meet 'cha, ma'am." He tipped his hat and hoped his hand wasn't shaking too much. "I've run clean out of supplies..."

"I've been expectin' you for days, Caleb. Why the whole town's been talkin' about you. 'Where'd the stranger come from? Why's he here?'"

Caleb swallowed, hard. *What if she knew?*

"I've sold more coffee since you showed up than I did on election day," she bubbled. "That long-sleeved blue shirt of yours is what's done it. The inmates at the prison wear 'em, you know—and they're always escapin'."

Inmates? Caleb's neck prickled, and he fumbled with the buttons on his right cuff.

"Why the day you pulled up beside the gas pump with a license plate so rusty even the sheriff couldn't read it, Darlene came in here a screamin', 'Belva! You better sleep with that pistol in your hand. There's a newcomer in town—and he's lookin' for sumpin'.'"

Caleb felt a rush of heat. He glanced uneasily at the dust-covered shelves, the lightbulb dangling from the ceiling on a long black wire, chicken feed leaking from burlap sacks and a poster taped on the window—COME PICK CHILES WITH JESUS—July 12. He reached into his pocket. "I got my list here somewhere," he began. "I need beans... and spam... "

Belva smiled and stepped closer, the huge dog clinging to her hip like Velcro. "Let's you'n me get acquainted first, luv." Her warm hand touched his arm, his right arm. "Coffee's ready—and I won't take no for an answer."

"That's right kind of you, ma'am."

"And this is 'Dog'," she said. "Someone dumped him off 'bout twelve years ago. Nothin' but ticks and bones, poor thing. Couldn't think of a name for him." She stroked the animal's massive head as it gazed at her with great, loving brown eyes while its tail swayed like a windmill rudder in an uncertain breeze.

"Nice boy," said Caleb, not too sure of the dog's intent when it pressed its nose against the lingering scent of jerky in his pocket. "He... he's nearly as big as a steer."

"He's my *man*," said Belva, "and a darn sight better than any of them two-legged, beard bristlin' ol' rakes walkin' around. Never met one yet that's worth what my Dog is. Lazy good-fer-nothin' bums, every last one of them! Interested in three things—booze, sex, and my money!"

Caleb felt Dog's hot breath seep through his jeans. His gut tightened—until Belva guided him gently into the adjoining café where five mismatched tables and chairs stood at random amid smells of unwashed ashtrays, yesterday's soup and today's baking. A ribbon of sunlight slipped between grease-laden curtains, accenting layers of peeling wallpaper and unclaimed hats on a bullhorn rack.

"Set yourself down, Caleb." Belva pointed a plump finger to a booth beside the jukebox. Before he sat, Caleb looked through the glass at the turntable. A 45 rpm record had stopped half way through "You Ain't Nothin' But a Hound Dog." *When?* he wondered. *How long ago?*

While Belva busied herself at the stove, he ran calloused fingertips over names and initials carved into the

Formica-topped table: PACO LOVES ROSA. ANDY LOVES _____. The name had been gouged out. ROBERTO AND THERESA FOREVER. ANDY LOVES _____. And again, the name was gone. *Good old Andy,* he mused, *loved 'em and left 'em.*

Belva placed two steaming mugs of hot coffee and a plate of fresh donuts on the table before squeezing herself into the too-tight space on the opposite bench.

"Smells good... ma'am."

"Oh, call me Belva," she wheezed, rooting through the shoulder area of her voluminous smock. Then hooking her thumbs under her brassiere straps, she lifted and dropped her pendulous breasts on the table where they quivered dangerously against the cream pitcher. "Help yourself, Caleb."

His face burned, but he reached for the cream that rippled and sloshed in the pitcher—*just like the water in the bucket had—so many years ago.*

"Sugar too?"

"Yes, ma'am... Belva." Again his gaze fell on the gouged-out names that had once been part of ANDY LOVES. "How... how long have you lived in Arizona?" he asked.

"Ever since me and Andy took up together. He thought he'd strike it rich down here. I should'a stayed in Vegas with Archie, or waited for Barney to come back from the war. They knew how to make a girl happy." She sighed. "But Andy promised me a gold mine and pretty dresses and a jewelry store. I got the store. No jewels though. The bottle killed him before he found the gold."

"I'm right sorry."

"Would you believe, after he died, I had to sleep with the little pearl-handled derringer he gave me under my pillow? I wasn't safe. Goddam men. But they all called me—beautiful."

Caleb peered over the rim of his cup and caught the familiar gesture, the sweeping away of a wisp of hair no longer there. He wished he could tell her that she was safe now... that she didn't have to worry.

"Folks stopped comin' to the store after Andy died," Belva went on, "so I paid a couple a' wets to build the café. I'm a good cook and I figured everyone's got to eat. And," her voice softened, "I had to have someone to talk to."

"That was a right smart thing to do."

"Smart! Me?" She laughed. "That's a first! I was always the 'beautiful dumb blonde', except to Andy. For him I was 'stupid woman,' especially when he was drinking."

"But it takes a good woman to run a store all alone, Belva. Folks need you bad way out here. Like I said, you gotta be smart to do what you've done, real smart." It was nice talking with her, getting acquainted.

For a moment she squinted at him. "You're not a cowboy, are you, Caleb?"

He shook his head. "Railroad. Track repairman till the darn cancer got my ears. Too many years in... the sun. The doc told me to wear a cowboy hat and... long sleeves. But how'd *you* guess?"

"You still got all your fingers. Ol' cowboys are missin' at least two, always gettin' 'em pinched off dallyin' that rope around the saddle horn when they go to lasso a steer."

Caleb looked at his stiff, swollen knuckles with new respect. "You won't tell anyone, will you?"

"Course not." It was then he noticed her eyes, larger than he remembered—two huge, blue pools of life and pain, veiled behind dust-tinted glasses.

"What brings you to Earhart, Caleb?"

"Needed a job. That Social Security's kind of a death warrant, you know. A man's gotta work. Makes the days go faster. It's just... the nights."

"You need a dog," she said brightly as Caleb glanced at Dog asleep at his feet with one eye open. "Now that's *real* company. They don't argue. They don't cheat on you, and they love you—no matter what."

"But they can't talk with you."

"Makes no difference." Belva stroked Dog's head. "After a while you can tell what they're thinkin', and they *know* what you're thinkin'. That's communicatin', sorta like me and Andy those last seven years. We scarcely talked. Didn't have to. When I picked up those dirty socks, he knew I hated him, and when I had to lay on my belly to fish the beer cans and empty whiskey bottles from under the couch I wanted to kill him. He knew that, too." She looked at Caleb then. "But at night, all alone, out here in the middle of the desert, at least there was someone to touch... to hold. Then he was gone. Thank god Dog showed up." Dog's tail thumped softly on the linoleum floor, his own personal reply.

Caleb didn't forget that chat with Belva, Dog at his feet, ANDY at his elbow, and it wasn't long before he made another visit, and another. Every time he promised himself... *today's the day I'm going to tell her*. But it never happened. He told stories instead, about the Navy, his railroad job, a bad marriage. She talked about her rosebushes, her paintings, and broken dreams. Then, when winter came, and the full weight of loneliness descended on him in his cold metal shack by the cotton gin, he listened to the relentless wind and thought about Belva in her cozy café near the crossroads... and he thought about another time, an earlier intersection in life, so many years ago. *Would she remember?*

One evening Caleb arrived at the café later than usual. A CLOSED sign hung on the door. He knocked. Belva's small apartment was attached to the rear of the building, and it seemed too long before the door opened.

"Oh, Caleb. Come in. I'm so glad you're here," she said. "Dog's ailin' somethin' fierce. He's trying to lie down by the woodstove. But he can't. Maybe you can help him!"

"Sure, Bel." It had been so easy to drop the 'va'.

"You can rub some liniment on his hip for me. Maybe then ..."

Caleb followed her inside. He'd had his hair cut but she didn't notice. *Too worried about Dog*, he supposed. Besides, it was dark inside. She disappeared into the kitchen. "OK if I turn on a light?" he asked.

"Oh, sure. I musta' forgot."

He groped beneath a lampshade and pulled the chain. Light flooded the tiny living quarters, and signs that Belva lived here were everywhere. Paintings of flowers softened the walls. Needlepoint pillows littered the couch. Books, covered in dust, lay everywhere; crammed onto shelves, shoved under chairs, stacked in piles on the floor. Yet she'd never mentioned she liked to read. Then he saw the eyeglasses.

Dozens of pairs cluttered the top of a cedar chest. Mail-order catalogues boasting new styles lay open... forgotten. And, on a small wooden table beside an overstuffed chair, black-and-white photos told the story of a beautiful young woman in latex bathing suits, evening gowns, and rhinestone tiaras. In every pose, her blonde hair whirled about her shoulders and fell seductively over one eye.

One picture, framed in tarnished silver, caught his attention. His hands trembled when he picked it up and blew dust from the glass. He held it under the light. The girl was surrounded by a football team, eleven husky young men, helmets under their arms, with "Las Vegas High School" stamped across the front of their uniforms. Kneeling on the ground in front of them all, another young boy clutched the water bucket. He wore a white shirt and dark trousers, and his eyes brimmed with a crush only a fourteen-year-old can feel for a beauty queen.

"Can you believe that was me?" Belva's fingers caressed the frame.

"Beautiful," he said, "just like a movie star."

"You're sweet. Right after that picture, Andy and I eloped. I wasn't quite sixteen. That's Andy with his arm around me, and Archie, and Leon... Harry... Chuck... and Barney's the one with the dimple in his chin."

Caleb remembered them all, Barney especially, the wrong friend at the wrong time... with a gun. "And who's the little guy with the bucket?" he asked.

97

Belva shook her head. "I guess I never noticed him."

Dog's whimpering grew louder. He turned in circles. And groaned. "Just listen to him, Caleb. Poor baby. Maybe you can lay him on his side and rub this liniment on his hip. You can get down on the floor lots easier than me." She smiled, the same smile that had bewitched them all so many years before.

"You sure he won't mind?"

"Oh no. He likes you, Caleb. He likes you a lot."

Kneeling was torture, a painful reminder of all those years on his knees, praying, doing time in the Yuma cell. But he got the old dog down and rubbed the salve in. Finally, the animal closed its eyes. Caleb, pleased at his success, forgot his own discomfort and passed the time reading titles on book covers surrounding him. "I never knew you liked to read so much, Bel."

"I always liked to read. And you know something funny? Nobody believed this dumb blonde could read."

"Now I know what to get you for Christmas," he said, amazed at the liniment's soothing warmth seeping into his own knuckles, deep into prison scars.

Caleb had supper with Belva that night and they talked about Dog. He would never get well. Caleb knew it. So did Belva.

"What am I going to do when he ...?"

"I'll find you another one." He meant it. "A nice little fella you can cuddle on your lap."

"I don't want another dog, Caleb." She turned away. "I want Dog. I need him. You don't understand."

A few weeks later he bought her a paperback romance. "I was going to wait 'til Christmas," he said shyly, "but that's six months off."

"Thank you," she murmured. "But I'd rather talk, Caleb. I can read when I'm alone." He felt a certain warmth in her words. Then he noticed a book he hadn't seen before. He picked it up and read the title aloud. "*What's in a Name?* Who got you this one?"

"Andy. He told me I'd know the truth about him if I looked up his name. So I did. It said Andy means 'manly'! He was so proud of that. Then I looked up some others. Archie means 'bold.' And Barney means 'noble.' Seems like I always picked real macho men!"

Caleb riffled through the pages. "I guess you already know that Belva means 'beautiful one'?"

"Yeah!" She blushed. "I looked that up first."

"And who were you looking up today?"

"Caleb," she answered softly.

"And what did it say? Handsome? Courageous?" He wanted to make her laugh, rekindle the smile he'd never forget. "Gallant maybe?" That's when he spotted the magnifying glass—shattered into a thousand pieces on the other side of her chair.

"Oh, Caleb, I don't know what it said! I... couldn't... see!" Her voice broke miserably. "I couldn't see the words at all."

Caleb stood up, the puppy love of another time replaced by an irresistible longing to put his arms around this woman who had lived alone with her secret and her dog for so many years. He wanted to comfort her, take care of her; but a piteous whine filled the room. A shuddering breath. Then silence. Dog was dead.

Caleb stayed with Belva that night. He urged her to

talk about the good times she and Dog had shared together. How every Sunday he walked her through the mesquite and the prickly pear to church in the trailer a half-mile away. She always arrived without a scratch, and she "couldn't see beyond her own feet." But nobody knew. She told him how the thwacking of his great tail against the bedpost had signaled the time of day, or drummed loudly just before dinner, or simply swung with devotion. Dog had been her eyes, her constant companion, her love, for fifteen years. Now he was gone.

The following morning Caleb built a coffin. By the time he started digging the temperature had soared to 112 degrees. Although the palo verdes cast welcome shade on the spot Belva had chosen, the earth was rock hard. The dig would take time, and since he was alone he stripped to the waist. It was safe—no one around to see the tattoo that stretched from his right shoulder to the elbow.

"She's beautiful. Just like a movie star." That's what he'd told the Philippine artist 45 years ago. *"Her blonde hair's got to whirl around her shoulders and hang over one eye. Her body? Well that's up to you."* The result? She was naked. She was gorgeous. He was the envy of the whole crew. And penned beneath the lavender robe at her feet … her name… BELVA.

Caleb shoveled the last bit of earth over the grave and, still on his knees, finished preparing a hole for the Tombstone rosebush. Too late he felt Belva kneel down beside him. "Plant it where Dog can smell it," she said. "Roses… were his favorite flower." Her breath caught. He felt soft, gentle fingers touch his arm. "That must have hurt," she murmured. "The war?"

He said nothing. He finished planting, put on his blue shirt and, taking Belva's hands in his own, he helped her up. For a moment he wondered about the smile flickering on her lips, but it was the tears that tugged at his heart. He pulled her hand through the crook of his arm and they walked in silence back to the store.

Some day, he thought, *when we get old, I'll tell her the truth about why I came to Earhart. I might even tell her about the tattoo—and those years between. And, when the time is right, I'll open that book called* What's in a Name? *and read aloud to her the passage that says, "Caleb means 'Dog'"....*

...and we can laugh together.

The End

Penny Porter lives and loves a writer's life. She has published stories in a wide range of national circulation magazines, including *Reader's Digest, Arizona Highways, American Heritage, Catholic Digest, Guideposts, Woman's Day, Honda*, etc. and her work has appeared in numerous international anthologies, world wide text books, and thirteen in *Chicken Soup for the Soul* books. The recipient of *Arizona Highway's* 2001 Silver Award for Excellence in Writing, Penny is the author of six books, most recently another collection of short stories, *Adobe Secrets*. A mother of six, grandmother of eight, great-grandmother of one, former teacher and school administrator, she is a frequent speaker at writers' workshops and former President of The Society of Southwestern Authors.

PennyPorter1@aol.com or write to Penny Porter, 7081 E. Calle Tabara, Tucson, AZ, 85750 for signed copies of *Adobe Secrets, Heartstrings and Tail-Tuggers or The Keymaker*.

Other books by Penny Porter

Heartstrings and Tail-Tuggers
The Biography of Eugene Gifford Grace
Howard's Monster
The Keymaker
Green Eggs and Sam

Anthologies:

The Rocking Chair Reader: Memories from the Attic
The Rocking Chair Reader: Family Gatherings
Keeping Christmas: Barbara Russell Chesser
Chicken Soup for the Soul: Christmas Treasury
Other Great Cat Stories: Joe L. Wheeler
Other Great Dog Stories: Joe L. Wheeler

www.ingramcontent.com/pod-product-compliance
Lightning Source LLC
Chambersburg PA
CBHW022121280326
41933CB00007B/492